LEADING
FROM
THE
FRONT

WINGS OF FIRE
An Autobiography

Respected Sp. Wahiji,

I & adore you for the contributions, what you have made to our Nation.

Greetings
A.P.J Abdul Kalam
21/6/2001

Recorded on book presented to Col. S. P. Wahi

Kalam accepts Wahi as his role model

DEHRADUN April 28: Bharat Ratna Dr. APJ Abdul Kalam, principal scientific advisor to Govt. of India sees former ONGC chairman Col. SP Wahi as his role model.

The startling revelations came off during the inaugural function of Fourth Petroleum Industry R&D Managers' Meet, with Dr. Kalam doing its opening.

Source - Himachal Times April 29, 2001

LEADING FROM THE FRONT

FROM ARMY TO CORPORATE WORLD

Col. S. P. Wahi

STERLING PUBLISHERS PRIVATE LIMITED

STERLING PUBLISHERS PRIVATE LIMITED
A-59 Okhla Industrial Area, Phase-II, New Delhi-110020.
Tel: 26387070, 26386209; Fax: 91-11-26383788
e-mail: sterlingpublishers@airtelbroadband.in
www.sterlingpublishers.com

Leading from the Front: From Army to Corporate World
© 2007, Col. S.P. Wahi
ISBN 978 81 207 3539 2
Reprint 2008

All rights are reserved. No part of this publication may be reproduced, stored in a retrieval system or transmitted, in any form or by any means, mechanical, photocopying, recording or otherwise, without prior written permission of the publisher.

PRINTED IN INDIA

Printed and Published by Sterling Publishers Pvt. Ltd., New Delhi-110020.

Dedicated
to

My late Mother
who inspired me with her fundamental values,
ethics and basic principles of management
for working with courage and in harmony with others.

Preface

I have ventured to pen down my experiences in the domain of management, due to repeated requests from my old colleagues and managers who have been attending workshops on leadership organised by me since my retirement from active work life in December 1989.

It is a story of a long journey, from a remote town khushab (now in Pakistan) which did not even have electricity, during my studies till matriculation in 1943. Now I understand it is a hub of nuclear facilities. To be associated with ONGC, a premier energy corporation, during my time a commission under the act of parliament, was a dream come true, after having worn Indian Army uniform for over 22 years and having worked with six public sector and government undertakings. Later as a consultant, I assisted many organisations both within the country and abroad. It is an experience of over 58 years, full of trials, tribulations, failures and successes.

My development in the younger days was inspired by my mother. The religious books of all faiths, she had made me read, the organized life she had ensured, as a role model of courage, commitment, time management and concern for the under privileged.

The hunger for knowledge and information was created during my training in the Indian Military Academy and EME training establishments. The study of autobiographies and biographies of great military, business and political leaders of different countries brought out important management principles, thoughts, values, abilities and characteristics required to be a good leader. The Chinese saying, based on the epic, Art of war by Sun Tzu, that "business is like a battle front" is so true in the present ruthless business environment.

The lessons learnt through my association with outstanding leaders, both junior and senior, from military and business gave me the courage, to experiment with success, in various aspects of management to achieve the objective of growth with stability and continuous improvement in productivity in a number of organizations.

Some astonishing data, on communication gaps, egos, group loyalties, lack of trust in each other, low morale, square pegs in round holes, uncontrolled growth in manpower in low technology and unrelated business, low emphasis on training particularly on leadership development, affecting the enterprise performance, was projected through 'attitude surveys' conducted in three different enterprises. Results of attitude surveys followed by meeting, a large cross section of the employees including the union leaders, reinforced my earlier experience that management of human resources has to have even higher priority from the management, than the organisation structure, systems and procedures, other functional and technological skills, to improve the performance of the enterprise. Human relation problems are the root cause of low performance of the enterprises.

Development of leadership at all levels, apart from managerial and functional skills of managers was identified as the priority concern. Leadership with an ability to inspire through character, courage, commitment, competence, concern for the people and the organization, ability to conceptualise and communicate. (7 Cs mantra). Leadership which believes in friendliness, fairness and firmness (3 F mantra) in its dealings and takes decisions only after taking into account the morale and motivation of the employees and money the bottom line of an organisation (3 M- mantra). The emphasis throughout my career has been to create a culture of development, mistake tolerance, healthy dissent, team performance using individual strengths and skills, collaborative approach, and providing dignity and respect to human resource at all levels.

The job rotation of managers in different functions, enable them to go up the ladder by proper understanding of the complexities and technologies involved in each function. No doubt a lot of commitment and hard work is required to understand different businesses and technologies. My own experience of working in different organisations with a variety of technologies did help to motivate others to follow the lead, without fear of failure. In spite of meticulous, succession planning, the whims and fancies of the political masters do play havoc with organisational growth and success. The stories of Cement Corporation of India and ONGC, in this regard are covered in some detail.

Having worked in almost every state of our great country, and visited most of the developed and a few developing countries, with different political shades, has reinforced my confidence in Indian talent both in the Public and Private sectors to excel and achieve international norms of performance. Much against the resistance, of some vested interests, major thrust was given to the development of indigenous industry with success.

The 20 year long-term plan of ONGC, which was widely disseminated, had almost created an industrial revolution for development of equipment and services hitherto imported, as indigenous industry could take long-term investment decisions based on the requirement projected by ONGC. The employees could see their own growth with the growth of the organisation, thus creating the desired motivation, morale and sense of belonging.

The heroic deeds in the face of danger during crisis did bring out the need of leadership, to be available in the theatre of operations during the crisis. This enables the operating people to have the desired support for prompt decision-making. A control room to document all decisions and the authority taking such decisions, does help to identify leaders with character and courage. The Sagar Vikas blow out details the dare devil actions motivated by leadership, and the courage to overrule the world renowned consultants to avoid colossal damage to the oil platform. The necessity to lead from the front is obvious.

The risks involved in conducting a type test on a 235 MW generator of Soviet design for the nuclear power plant, without reference to BHEL board and Soviet designers, brings out the ability to take decisions in the interest of the organisation without worrying about personal accountability, in case of a failure. The decision to conduct this test was held up for over a year, prior to my joining. The leaders have to have the broad shoulders, to be accountable for the results even when the powers have been delegated down the line.

It is the responsibility of the CEO particularly in the public sector enterprises, to insulate the organisation from unnecessary bureaucratic and political interference through 'diplomacy' and if required, through 'frontal attack'. The leader should be ready for personal sacrifice. The incidents narrated in chapters concerning BHEL, CCI. and ONGC would illustrate the issue in depth.

The role of aggressive communication and honest information sharing can be of great assistance in maintaining good performance even under difficult conditions. ONGC had almost uninterrupted operations even in hard duty operational areas in Assam, Tripura and Nagaland. The operational results in these areas, during the period 1981-89 bring out clearly the advantage of proper relationship through communication outside the organisation at various levels. During crisis situations, communication and courage of leadership are the most important factors for success. This is illustrated in some detail in the chapter relating to Sagar Vikas episode.

Strategies followed to achieve creditable performance in politically difficult environments bring out the confidence, courage and commitment required by the leadership to establish relationship and contacts at the working and political levels with sincerity and trust. The chapter – 'Relationship Management' covers the issue in detail.

ONGC was always ahead of times in projecting views to the government for action on oil sector reforms and issues connected with energy security.

It is a pity that some of the initiatives projected to the government for energy security, formulation of an integrated energy plan. Re-engineering of the oil sector, to have national integrated oil company, national gas grid, purchase of high technology companies in the developed world, did not get the desired attention, which these initiatives deserved. Some actions on some of these are now being forced by circumstances.

The Corporates are islands of prosperity with the poverty all around. They have a responsibility to meet the needs and environmental and safety concerns of the society within the guidelines of corporate governance. The role of women in this regard has been very creditable. How these responsibilities were met are covered in some detail.

Col. S. P. Wahi

Acknowledgements

My acknowledgement and thanks are due to all my colleagues and co-workers without whose combined talent, competence and contribution, the success stories of this book relating to Army, and other Public Sector Undertakings, could not have been possible. My erstwhile colleague Mr. R.Srinivasan, Dr. S.Ramanathan & I.A.Farooqi have helped to fill the missing links and gave the book much desired polish to be intelligible. Maj. Gen. K. B. Kapoor reviewed the manuscript and made useful comments.

I was inspired during my career by many colleagues both senior and junior. The list is exclusive and I can only mention a few due to limitation of space – Late Mr. Mantosh Sondhi, and Dr. V. Krishnamurti, Lt. Col. Balwant Singh AC, Late Brig. R.P.S Randhawa. Amongst those who inspired me, like to million others, are Mrs. Indira Gandhi and Mr. Rajiv Gandhi.

I have drawn heavily on the book 'Story of ONGC' by I.A.Farooqi a brilliant professional and a historian of great talent. I have also taken guidance from the work of another great professional and historian Dr. Vishwanath. Recently, I have read with great interest the book 'Up Stream India' of ONGC by the editorial team headed by consulting editor Mr. Raj Kanwar.

Mr. D. N. Awasthi and Mr. Kuldip Chandra, two eminent Geoscientists of ONGC gave me initial briefing on ONGC and the game of oil exploration. Dr. Reena Ramachandran, an eminent technocrat managed with distinction management services department, communication with internal and external environment, apart from scanning of economic and technological environment for future energy options.

There is no sustenance greater than the affection and support of ones' wife. In this regard Shobana was a great strength and help all through my career. She played a sterling role to manage welfare activities for the families of employees in the areas of our operations and established schools, polytechnics, vocational centres, hospitals, and conducted other social services for the community. She in addition played the parental role on my behalf for our three children Rakesh, Sunanda and Shalini. Rajeev Lal and Rajan Chauhan our sons-in-law, our children and grand children have provided love and support. Saloni along with Rakesh have provided motivation, help and support to complete this book.

Being the youngest of five brothers and three sisters, I received abundant love and support from them during my development in the formative years. They along with their married partners were of immense help and support.

The book could not have seen light of the day but for the immense help I received from Mr. Ashok Varma, Dr. Seema Sanghi and Mr. A.S. Soni. My gratitude is also due to Mr. J. C. Kapur, Mr. A. Kanwar Singh and Mr. S. K. Aditya for their tireless secretarial help.

Last but not the least my thanks are due to Sterling Publishers for bringing out this book under the dynamic leadership of Mr. S. K. Ghai and his team.

Col. S. P. Wahi

Contents

Preface   vii
Acknowledgements   xi

Foundation Years

1. Early Days   3
2. Days of Tribulation   17
3. On Foreign Deputation   26
4. The Instructional Years   28
5. Deputation to Heavy Vehicles Factory—Avadi   33
6. Back to the Army   37

Army to Corporate World

7. Beginning with the Corporate Phase—Bokaro Steel Plant   43
8. Leading BHEL (Haridwar) to New Heights   48
9. Strategic Planning—Cement Corporation of India   65

My Tryst with Black Gold

10. Oil – the Life Blood of Economy   79
11. Thrust on Oil Exploration   88
12. Business Is Like a Battle Front   102
13. Restructuring and Organisational Changes   121
14. Bombay High Oil Field—Canards   131

Effective Management

15. Leadership and Communication in Crisis Management (Sagar Vikas Blow-Out)   141
16. Relationship Management—Creating a Collaborative Environment   151

Performance Management: Consolidation of the Public Sector Oil Companies

17. Moving through the Bureaucratic 'Jungle'   159
18. Re-engineering the Oil Sector—Globalisation Imperative   171
19. Oil—Economic and Political Weapon   188
20. Vision for Tomorrow   195

Appendices

Appendix A	: DO No. 6/26/ CH. OIL & Natural Gas Commission, Tel Bhavan, Dehradun dated 20th October 1981	215
Appendix B	: Oil Field Deals; Bad When in Opposition, Good When in Power	222
Appendix C	: Relying Only on Oil Could Be Slippery	223
Appendix D	: Chairman's Call for Intensifying Exploratory Efforts	225
Appendix E	: Hon'ble President Visits ONGC R&D Institutes at Dehradun	226
Appendix F	: President Lauds ONGC's Achievements	227
Appendix G	: Performance Charts	228
Appendix H	: Certain Statistical Comparative Figures Mark the Growth of ONGC Between 1980-81 and 1989-90	230
Appendix I	: Petroleum – Past, Present and Future	232
Appendix J	: OPEC (Organisation of Oil Producing and Exporting Countries)	236
Appendix K	: The Inspiring Force Behind Sanghe Shakti	239
Appendix L	: ONGC Makes its Presence Felt Internationally	240
Appendix M	: DO No. 6/33/ CH/86 Oil & Natural Gas Commission, Tel Bhavan, Dehradun dated 1st August 1986	241
Appendix N	: The Cauvery Basin	256
Appendix O	: Minister All Praise for the New Vocational Centre	257
Appendix P	: ONGC Contributes Towards Beautification of Dehradun	258
Appendix Q	: OAPEC Delegation Visits Dehradun	259
Appendix R	: ONGC – The Colonel's 2-Front War	260
Appendix S1-S2	: Advisory Council on Exploration Strategy	262
Appendix T1-T2	: Prime Minister Lauds the Efforts of ONGC	264
Appendix U	: ONGC's Strategy for Environmental Protection	266

Foundation Years

1

Early Days

The dream to be an engineer in uniform came very early in life. The untimely death of my father in 1937, when I was just eight years old, created an uncertain environment, in which I was to grow. I was the youngest of five brothers and three sisters. The oldest brother was an engineer in the railways; the second had finished his engineering from Thomson Engineering College Roorkee, an elite institution for civil engineering, and was settling down in the Punjab Government, Department of Irrigation; the third had taken a degree in industrial chemistry from Banaras Hindu University; and the fourth was doing his graduation in commerce. My eldest sister was married and other two were three and five years older than me.

My mother, the two younger sisters and I moved to Lahore. After a year, circumstances forced us to move to Khushab, our hometown (now in Pakistan, about 210 km west of Lahore). Khushab was a trading centre. It was located on the banks of River Jhelum. The movement of goods was generally by boats and camels. In those days, Khushab did not even have electricity. We had to study under the light of kerosene lamps. Life was simple but rough and tough. One had to carry a knife and a flexible hard-hitting whip for self-defence. The only entertainment was through mobile cinemas and village fairs. One had plenty of time to indulge in rural sports and community social events—kabaddi, wrestling, hunting, horse riding, *Ramlila* and dramas. My father had bought me a horse, which had to be sold after his death, but my interest in horse riding remained. The circumstances and the environment fuelled my desire to equip myself with skills, abilities and knowledge. Therefore, I tried my hand at every activity or task which came my way. I kept a lot to myself and spent plenty of time reading Urdu novels and books.

My mother was a great source of inspiration, being a pious and strong human being. She was a strict disciplinarian, with a very planned and organised life. She would punctually get up at 4 a.m. and after puja (prayers) do a bit of spinning of yarn. She was a follower of Mahatma Gandhi in this respect. We were made to get up early in the morning and to study before going to school. In the evening we had to be home for aarti (evening prayers), eat, study and sleep by 9.30 p.m. This habit of early to bed and early to rise has remained with me till date. So has been the habit of strict time management, which paid good dividends later in life.

She demonstrated and inculcated in us a strong character, values, and courage to face problems with confidence. She would read to us from the *Gita* and forced us to read the *Ramayana* and *Mahabharata*. She had a photographic memory, very clear objectives and a strong will for action. She was a visionary and was a role model for all of us in the family. With very limited resources at her disposal after my father's death, she ensured that we had a settled life.

Women from nearby houses used to come to her with their problems for advice and to resolve conflicts. She believed in sacrificing for the poor and the underprivileged. One cold winter morning, a scantily clad sadhu came for help. My mother asked my nephew, who had also joined us at Khushab for studies, to bring one of the two blankets which we used while studying, to be given to the sadhu. He brought both. She however gave one to the sadhu. Next day the sadhu came with more sadhus and requested for the second one too. My mother gave it away as well. Both my nephew and I had to do without blankets till we could afford to buy replacements. Such incidents of generosity on her part had a great impact on our attitudes and values. She used to remind us that the greatest happiness is to do good to others.

Our family used to travel during school holidays to Dalbandin (near Quetta now in Pakistan) where my eldest brother was posted in the Railways. There was a train only once a week from Quetta to Dalbandin and on to Zahedan (a town on the Iran border). Being the only male member in the entourage, I had the responsibility of escorting my mother and sisters from Khushab by train to Dalbandin. Responsibility came very early in life. We enjoyed free rail passes, as dependents of my brother. Those were the days of the Second World War. Movement of troops engaged in war games, and watching the railway engineers in action to maintain and repair railway tracks and equipments in the harsh desert environment of Baluchistan (now in Pakistan) was a great motivation for me to be an engineer in uniform. This further reinforced my commitment to work hard and realise my dream.

> This attitude to rebel against injustice remained within me.

My brothers were very kind to provide me with all the material and financial support, but I still missed the emotional support of my father and remained a lonely child. My sisters recall that after the death of my father, when I was a little over eight years, during one of the family get-togethers at my brother's house, I had threatened to leave the house over an argument. I was asked to leave only with the clothes I was wearing. I rushed outside the house and sat under a tree, as I felt I had received an unfair treatment. The family members had treated it as an amusing incident, but were no doubt worried about my next move. I came back and asked to be allowed to take at least my flute with me. It was only then that my brother realised the seriousness of the matter. He and everyone in the family embraced me and consoled me and conceded that I was right. This attitude to rebel against injustice remained within me throughout my working life. Lady luck continued to favour me, as I did not visibly suffer on account of this attitude of mine.

After matriculation, I had joined the Government College Shahpur (about 15 km from Khushab). Most of the students were from affluent landlord families, who had little interest in studies. The environment for studies was missing. In view of this, I was forced to shift to Lahore during the middle of the first year. I was able to get admission in DAV College and got accommodation in Dev Samaj, a private hostel. In the hostel, I came across students from various colleges and got a good taste of cultural backgrounds of students from various strata of society. Apart from studies, I also took lessons in Yoga and played hockey, a game in which I had excelled in school.

Engineering Education

The competition to get into engineering colleges was tough. I got admission in the Engineering College, Banaras Hindu University. There were 158 students in the first year, with an excellent mix of brilliant students from every state of the country. Apart from the fact that each student was good in academics, they were talented and participated actively in sports, music, art and other cultural activities.

Banaras Hindu University had colleges in every branch of education and had excellent facilities for sports, music and art. The university was set up by a great visionary – Mahamana Pandit Madan Mohan Malaviya in the year 1920. The architecture of the buildings and hostels of various colleges is unique. Gita discourses of Sir S. Radhakrishnan (later became the President of India), who was then the Vice-Chancellor of Banaras Hindu University, were a source of great inspiration to the students and had great spiritual value.

The engineering college had outstanding professors. Every faculty member had some distinctive characteristics and strengths. Two stalwarts had a great impact on my character building. One was Maj. M.C. Pande, a professor in electrical engineering, a company commander in UOTC/NCC and warden of Rajputana Hostel, where I had spent four years under his direct supervision and guidance. I was also a member of UOTC/NCC for all the four years, in view of my childhood dream of donning the uniform. Maj. Pande was a father figure and showed great interest in every ward. He was available for assistance and advice at all times. He knew the strengths and weaknesses of each one of us and had a good sense of humour. He kept an eye on each one of us and kept us out of mischief. The other was Prof. M. Sengupta, who was professor of electrical engineering and the principal of the College. He was an intellectual giant and a very pleasant personality. He taught us electrical machine design in the final year. Two other students and I had taken up this special subject. In view of his administrative responsibilities as the principal, he found very little time to teach only three students. He would call us to his house for teaching on holidays. He would also entertain us with delectable Bengali sweets.

Apart from being in NCC, I played hockey, took part in weight lifting and athletics, and won many college and university prizes. Summer vacations were spent in practical training in factories of Modi Industries at Modi Nagar where one of my elder brothers was employed. In the third year, I was also deputed for a few months to Bombay Electric Supply and Transport Corporation (BEST) for practical training.

Activities Prior to Joining the Army

After graduating in both mechanical and electrical engineering, I appeared before the Army Service Selection Board at Meerut for induction into the army. While awaiting the results, I worked for a short while in Assam Rail Link Project. This difficult project was headed by one of the most brilliant engineer administrators, Sardar Karnail Singh. I had watched with great admiration the building of the railway bridge over River Teesta near Siliguri—a rare engineering feat, accomplished by railway engineers. From there I moved to Joginder Nagar Powerhouse (hydroelectric power) in Himachal Pradesh for training along with a few other engineer trainees, who were regularly applying for jobs. My mind was totally for the army and therefore made no effort to look for alternative jobs. However, on persuasion from fellow trainees, I also sent my bio-data to Birla Engineering College at Pilani. Only on the basis of my bio-data, the college sent me a telegram to join forthwith. I was young and made the mistake of joining the staff of Birla Engineering College, without checking the terms and conditions of employment. The Principal, Mr Lakshmi Narain was an outstanding administrator, I had ever come across. I had to work along with the professor of electrical engineering and assist the professor of mechanical machine design as well. It was a great feeling to be teaching engineering when I was hardly 20 years old. Some of the students were earlier my class fellows in Lahore.

Through a telegram, I was asked to join the Indian Military Academy Dehradun in January 1950. I requested the principal to relieve me. To motivate me to stay with the college, he offered me all types of incentives, including a commission in NCC, deputation to UK for higher studies. He also mentioned that I was bound by a three months' notice period. This was an unpleasant surprise. I had signed no such bond. My dream of being in uniform nearly got shattered. Maj. Subramaniam who was the Vice Principal, helped me to literally escape. It was an incident I would like to forget because of my deep regard and respect for the principal.

Strong Foundation for Leadership

The Army emphasises continuous training and development of its officers, to enable them achieve excellence in professional/functional, managerial skills and leadership abilities, to win battles and wars. Offices are even trained to help civil administration in times of crisis. The troops below officer rank also receive continuous training. They are encouraged through incentives for promotion to upgrade their educational qualifications. They also have to undergo periodic professional courses to upgrade skills and improve abilities for planning and supervision.

The army has many different units (arms) such as infantry, armour, engineers (sappers), artillery, signals and services such as Army Service Corps (Supplies and Transport), Ordnance Corps, Corps of Electronics and Mechanical Engineering, etc. Basic professional, managerial and leadership training is imparted in the Indian Military Academy for gentleman cadets inducted for regular commission in the army. The engineering graduates and cadets on completion of their training period of three years at the National Defence Academy at Pune spend only one year in training in the Indian Military Academy, whereas the other graduate cadets spend two years. After Commission each cadet is allotted an arm/service depending on his choice and the assessed potential. Each arm and service has its own specialist training establishments where training is imparted to the officers on a continuous basis over the service period, to upgrade professional/managerial competence and leadership abilities. There are common training establishments for all arms and services of the army as well as all the three services (army, air force, navy) where training is imparted for the positions of higher command and joint integrated functioning.

Being an engineering graduate I underwent training only for a year at the Indian Military Academy where apart from military training various academic subjects such as economics, military history, human psychology, art-of-war, tactical and strategic planning are also covered. Sports, swimming, athletics and horse riding received high priority along with use of weapons. To improve the bearing, ability to walk in step with each other and create a sense of instinctive obedience, a lot of training on marching was imparted on the drill square. Due importance was given to the uniform including the 'spit and polish'. The drill instructors motivated one to walk with the pride of a peacock. The band helped to add pomp and show which created a great sense of enthusiasm and pride. Through many outdoor tactical exercises, route marches and camps, under different adverse climatic and physical conditions, the qualities of leadership and team working, interpersonal relations were strengthened apart from an increase in the physical stamina. Competitive spirit was developed through various inter platoon, inter company and inter battalion competitions. The gentleman cadets (GCs) as we were called, remained under constant supervision to ensure progress and growth. Periodic feedback was given to the cadets which also covered results of mutual assessment by cadets of each other, on qualities of leadership and friendship.

> **!** Army emphasises continuous training and development of its officers, to enable them to achieve excellence... **!**

The commandant and other instructors in the Academy were outstanding officers. Most of them had actual experience in military operations. They acted as role models in professional accomplishments, social behaviour and attitudes. Gen. K.S. Thimmaya (one of the greatest generals) was our commandant. The company and battalion commanders were highly decorated officers, some of them had won decorations during the Second World War and operations in Kashmir. The wives of the officers also played a very important role in improving the quality of life of cadets, by interacting with them during social functions.

Leadership Development

It was drilled into the cadets that 'the honour and safety of the country came first, always and every time, the welfare, safety and comfort of people under your charge came next and your own self came last, always and every time'. Through lectures, group discussions and study of military history and campaigns, we were made to understand that leadership is the main factor in achieving objectives and success. It came out loud and clear that leadership is the ability to influence/inspire a group of people to move willingly and enthusiastically to achieve a common group objective in a synergetic manner. One had to know one's people well, understand their dreams, sentiments, sensitivities, problems on and off duty; only then one could expect to get their unconditional commitment. The soldiers have to be ready to make the supreme sacrifice of their lives in battle. This can only be achieved through training and most importantly if they trust the leadership. The development of leadership abilities therefore is given very high priority during training.

Leaders have to know their people well to inspire and motivate them to put in their best. A deep knowledge of human psychology is a must for the leader to keep the morale of the people high and motivate them. A leader has to have empathy and strong emotional intelligence. Leaders have to have genuine concern for the welfare of the people under their command and their families during service life and also after retirement. The British officers had done considerable research on the characteristics of Indian soldiers from different parts of the country and had documented the same.

Leaders have to influence the command through character, courage, commitment, concern for the people, competence in professional skills, ability to conceptualise, plan strategies and tactics and communicate effectively. High standard of physical fitness was essential to be a role model.

The cadets were made to do every task thoroughly, even polishing of shoes, till a minimum standard was achieved. This also brought out the value of dignity of labour.

The routine in the Academy was so tight that one had to be on the run all the time and be physically fit and mentally alert. The emphasis on time management and punctuality created the desired sense of urgency in dealing with professional tasks. This characteristic got built into normal life, as well, with great benefit to ensure an organised life. Ample opportunities were available to develop hobbies and take part in cultural activities. Periodic tests for physical fitness, endurance, academic and professional subjects, created the motivation to work hard. The philosophy was that any one who ceases to improve ceases to be good.

Strategies and tactical planning was given high priority, so also understanding the need for long-term vision, unplanned contingencies, impact of weather, and other hostile environmental considerations for execution of plans in unknown enemy territory. It was emphasised over and over again, to appreciate the strength of the enemy and not to underestimate his potential, to avoid unpleasant surprises. The need for obtaining detailed information about the enemy's characteristics, his future plans, equipment and weapons, was appreciated through various tools. In this context, it is worth mentioning that the study of *Art of War* by Sun Tzu brought certain principles of war into focus which helped me while dealing with competitors in industrial life later in my career.

> Leaders have to know their people well to inspire and motivate them to put in their best.

Team Building

Route marches and camps for tactical training, taught one to live together on each other's strengths. Some cadets had the flair for cooking food, some had the physical strength to dig trenches and carry load and heavy equipment during marches. Some did not have the stamina to run for long distances and some like me had the weakness to enjoy sleeping and did not like sentry duties at night. We learned to supplement and complement each other's efforts. No one ever let down any colleague. One learned to live on faith and trust. The common uniform created a sense of belonging to the group and ensured removal of complexes.

During training, the need for good human relations, when seniors have to forget that they are senior and juniors have always to remember that they are junior, was realised. The respect of juniors has to be earned through professional competence, proper attitudes and personal example. A lot of emphasis was given and training imparted on the art of communication, both verbal and written. The need to keep each member of the group informed of the challenges ahead, and to tap the innovative and creative minds of junior colleagues was always stressed. In the army units, particularly in the field, it is a laid down practice to communicate with the troops in the evening through roll calls, about the latest news and challenges and get their feedback. The need to keep live interaction with the outside environment was stressed to get cooperation from all concerned, including civil administration, to meet the objectives.

There was constant interaction between the junior and senior cadets and a lot of inputs on traditions, behaviour and discipline were received from the senior cadets apart from some bullying and ragging in the initial stages. This helped one to get rid of inhibitions, complexes and sensitivities. It enabled one to face life with optimism and courage.

Towards the end of the training, each cadet was interviewed by a board of officers headed by commandant Gen. K.S. Thimmaya. In spite of a lot of persuasion by the commandant to opt for joining the Corps of Engineers, I persisted in opting for Corps of Electrical and Mechanical Engineers (EME) and was duly commissioned on 11 December 1950 in a very impressive parade and associated functions. The salute was taken by the Army Chief General (later Field Marshal) K.M. Cariappa. Being an engineering graduate, I was given two years seniority and my effective date of commissioning was antedated to 11 December 1948.

Systems and Procedures—Focussing on the Objectives

I had two month's training in Organisation and Methods at the EME School at Kirkee (Pune). A number of brilliant senior officers briefed us. They gave some informal tips as well. It was emphasised that in addition to what was learnt in the Indian Military Academy, for every activity a drill and procedure had to be laid down and followed. Even for loading transport for movement from one location to another, loading tables were laid down in detail as to where a particular item of equipment/article would be kept in the vehicle. It was however, emphasised that to achieve the main objective during the operations or while posted in field formations (where hostilities could start at a short notice), one need not subordinate oneself to procedure but use initiative and innovation to achieve the objective. One only needs to focus on the objective. This focus on objective remained with me throughout my working life. This enabled me to take lot of personal risks in the public sector Corporate World to ensure excellence in performance.

Out of the 24 of us who had passed out as regular commissioned officers in the third graduate course, I was the only one who was posted to an active field formation. My colleagues sympathised with me as it was a hard field posting. I look back with great satisfaction that I got the right exposure early in service. I was posted to a workshop being raised in Baramula (Jammu and Kashmir) under the command of a captain who had risen from the ranks and was not a qualified engineer. He was in his late forties. On the very first day he asked me to take charge of all activities, and delegated the powers, but advised me to get back to him in case I ran into any problems. I took him literally and went about managing the work and the troops with great enthusiasm, sense of urgency and speed. In the process I made a number of mistakes, some very serious. He never reprimanded or admonished me, but was able to set things right because of his long experience. This was true delegation. I learnt early in life that one can delegate responsibility and authority but never accountability; it was a very useful management principle, which guided me right through my career. I also learnt not to underestimate the competence of people who have risen from the ranks and are less educated.

> I learnt early in life that one can delegate responsibility and authority but never accountability...

Management of highly trained technical manpower in uniform was a great challenge. Living with them during exercises and being with them from PT/Parade in the morning, to repairing and maintenance of equipment was an extremely valuable experience. I had to visit them during their meals to ensure that proper quality and tasty food was being served, inspect the barracks/tents where troops stayed to ensure cleanliness, neatness and uniformity. My identification with the troops in all their activities gave them the pride in their work. One could see a smile on their faces even while working under difficult conditions. One learnt to listen to their individual problems and make sincere efforts to solve them. One also came in direct contact with troops and officers of other arms and services who were guarding the Cease Fire Line (CFL) in the Uri sector of Jammu and Kashmir. This was a unique opportunity to understand the concerns and inspirations of the soldiers and to enjoy their sense of humour and natural deference to competent officers, which is so necessary to build efficient teams.

The junior commissioned officers, non-commissioned officers, technicians and soldiers were from different states of the country. Apart from their professional competence, they were very accomplished in sports, cultural activities and arts. It was a pleasure to integrate individuals from diverse backgrounds and skills into a close-knit team to provide the desired repair cover for a variety of equipments in use in the Infantry Division guarding the LOC with Pakistan. The morale of the personnel was kept high by organising sports and cultural activities and giving attention and support for personal problems of individuals.

Director EME, Brig. Worthington (a British officer) made a visit to Srinagar. I also got invited to the dinner organised in his honour. To the horror of other senior officers present, I had asked him, "Sir what is the criteria for selecting officers for training abroad. We are struggling in the field and officers of our seniority are being picked up from peace stations for training abroad". He smiled but had no answer.

As I progressed in life, it became clear that individuals who had to achieve quantifiable results in a limited time had more onerous tasks than those occupying staff positions either in field or peace stations. Therefore, performance assessment of individuals had essentially to take this aspect into account. The army does classify officers into those capable of command/leadership positions and those fit for growth only in staff (advisory). Army officers for promotion beyond the rank of Major have to show adequate performance in independent appointments entrusted to them. Officers are also assessed for growth in staff or command appointments. Those in command positions have to show a lot of courage to take quick decisions and exhibit initiative. They have to have leadership abilities. They are the ones who will go up the ladder to the highest positions. During my service with corporates, I always had a soft corner for those working in the field/shop floor of factories, undertaking quantifiable results through initiative and commitment, exhibiting leadership abilities.

Professional Training

After a tenure of about 15 months in the field, I was posted to the EME School Kirkee for a training course on maintenance and repair of vehicles (jeeps, trucks) and armoured fighting vehicles. In the same course we had a mix of officers of different ranks from lieutenant (myself) to lieutenant colonel. On completion of the course, based on my performance I was recommended to be an instructor in the same institution. Army headquarters had planned my attachment with Ordnance Factory Kanpur for a year to be trained in the manufacturing of guns and other equipment. It was ensured that I was posted to a unit to pick up my rank of a captain, before proceeding on further training. I was accordingly posted on promotion as captain to a transport workshop company with ASC Regiment in the Patiala Brigade at Yole Camp (Dist. Kangra) in Himachal Pradesh. I had the very unusual experience of maintaining Harley Davidson motorcycles which were off road, as these were not in use in other army formations and did not have spares backup. Being a motorcycle racing enthusiast, I used innovative methods to repair these motorcycles and enjoyed using them during my stay in Yole. This place was a war camp for prisoners during Second World War. One could still see some beautiful paintings painted by Italian prisoners of war, on the walls of the barracks.

Orientation to Industry

I was posted to Ordnance Factory, Kanpur after a three months' stay at Yole. It was here that I received practical exposure to manufacturing processes and the pleasure of working shoulder to shoulder with workers. Perhaps the interest shown by a young officer motivated the workers to teach their skills. It was here that I received my first exposure to the mechanics of an industrial undertaking and got acquainted with the strength of the workers and the weakness of the management system.

The executives/managers of the Ordnance Factory were rarely seen on the shop floor. They remained out of touch with the practical problems and realities on the shop floor. There was a system of bureaucratic management, resulting in delays in decision-making. Also a culture of confrontation instead of collaboration with the workers was prevalent which caused low productivity. This made me realise the need for a new approach—the application of Human Relation as practised in the services to an industrial situation – an approach which was useful in my subsequent career in Heavy Vehicles Factory, Bokaro Steel Ltd., BHEL (Haridwar),

Cement Corporation, Bharat Opthalmic Glass Ltd. and Oil and Natural Gas Commission.

During my training with the Ordnance Factory, I stayed in the Army Ordnance Officers Mess. My interaction with Army Ordnance officers of various ranks helped to establish a close friendship with them which proved useful for cooperation in later service life. I also joined the Kanpur Club and enjoyed an excellent social life with members of the Club from various walks of life. I was also able to go hunting around Kanpur and had a fair amount of horse riding.

2

Days of Tribulation

After the completion of my training at Kanpur, I was posted as an instructor in EME School Trimulgery, Secunderabad. With this started a story of administrative mess and harassment which I faced with courage, patience and perseverance and which taught me to empathise with the problems of others.

I reached the EME School (Trimulgery, Secunderabad) and, was told that my posting stood cancelled. By mistake I was posted as an instructor in Telecommunication Wing instead of the Vehicles Wing where there was no vacancy. The commandant of EME School tried his best to retain me but army headquarters posted me to the elite 5 Armoured Workshop which at that time was with the First Armoured Division at Jalandar.

I arrived at Jalandar one evening and stayed in the Armoured Officers Mess. In the mess, I met one major of the EME who after learning that I had to report to the 5 Armoured Workshop next day, did not say a word. Next morning I reported to Maj. B.N.Malhotra, the Officer commanding the 5 Armoured Workshop. He asked me to accompany him to commander EME headquarters where he introduced me to Major Officiating CEME whom I had already met the previous evening. He said, "Captain Wahi either you drop in your rank from Captain to Lieutenant or proceed on leave". According to him the captain whom I had to replace was senior to me and his posting order had not been received. He was aware that there was a captain junior to me in the workshop who had to drop his rank by my joining but he was his friend. I said, "Sir, do you lay down army orders/instructions or the army headquarters'? After hearing this, Maj. Malhotra left the room as he sensed from my tone that a serious disciplinary issue could get created.

The major asked me to sit down. He then requested me to go on leave to save demotion of the Captain who was junior to me but was his friend. I agreed but asked him to give it in writing. I proceeded on leave and on reporting back after a month, I was given another posting order to report to a Transport Workshop Company (ASC) at Ferozepur. I thought I was being harassed and humiliated. I, therefore, refused to move from Jalandar and was marched before the Brigadier EME Western Command who was on a visit to Jalandar. On his enquiry about my problem, I complained that despite having done well in a training programme, I was being thrown around like a 'shuttlecock'. I expressed my intention to remain in Armoured Division. I was asked to report to the First Armoured Brigade where a major and a captain were authorised, and only one captain was posted who was drawing the officiating pay of a major. I reported at Kapurthala the same evening. This officer thought that I was senior to him and by my joining he would be losing his officiating pay. He had established a very good relationship with the brigade commander. In the morning I was told to report to the 18 Cavalry, as an advisor to Officer in-charge/Light Aid Detachment (LAD) at Jalandar. I went back to Jalandar and reported to the 18 Cavalry.

I was not aware of the full story when I had reported to the regiment. Even if I had known, I would have had no option as I had asked to be posted in the First Armoured Division. When I look back, this one decision of mine which was prompted by a vision to work with Armour, and to thwart the design of my detractor to get me posted to an ASC unit changed my whole future working life profile. Posting to the 18 Cavalry was given to me, possibly as a punishment as I had annoyed the Major. 18 Cavalry had not enjoyed a good reputation, with reference to the treatment meted to LAD (EME) officers. Hardly anyone had lasted more than a year with the regiment. They were all badly treated and some were humiliated. The one whom I had come to replace had been demoted from a captain to lieutenant.

While I was awaiting posting orders of Lt Nirmal Singh whom I had to replace, I was attached to the 5 Armoured Workshop to investigate the defect on diesel engines fitted on Sherman MK III Tanks. Col KSK Murthy, the CEME First Armoured Division visited the place where I was working with my hands and made some off hand remarks to which I reacted on an impulse and said something insulting. He looked to the other side and went away. He did not interact thereafter with me during my stay at Jalandar. I met him again when he was posted to army headquarters in 1959-60 as my direct boss. By then I had matured quite a bit. He was a brilliant engineer and a thorough gentleman. One of the best confidential reports I had ever got was from him. When I look back, I find that I have been lucky to work with outstanding leaders, with an odd exception, as even from them I had learnt the 'don'ts' of working.

> I have been lucky to work with outstanding leaders, ... from them I had learnt the 'don'ts' of working.

The future challenges slowly started unfolding when Lt Nirmal Singh, whom I was to replace gave me his pathetic story. He had been ill treated and humiliated. He appealed that I should relieve him immediately to save him from further embarrassment. I signed handing over papers without checking the stores and equipment. I went and reported to Maj. R. L. Chopra (He later retired as major general) who was acting as adjutant, that I had taken over the command of LAD. He said, "It is your funeral, if you have taken over without checking." Through anonymous letters, he had come to know about the shortages of equipment and stores in LAD. I did not reply but was confident of managing the situation.

A day after Lt Nirmal Singh left, officers from neighbouring EME units started ringing me up for the return of stores and equipment which Lt. Nirmal Singh had borrowed to complete deficiencies for handing over stores to me. Quite a few other serious discrepancies came to light which had to be sorted out later with the cooperation of the audit people. The morale of the LAD personnel was low as no Officer Commanding LAD was able to stay long enough with the regiment to raise it to the desired efficiency level.

I finally got posted at Jalandar. After having been made to shuttle from one place to another – from Kanpur to Secunderabad to Jalandar and to my home station on forced leave, then to Jalandar, Kapurthala and back to Jalandar, I put in a claim for travelling and daily allowance for all the moves to the Controller of Defence Accounts. His office agreed to pay only from Kanpur to Jalandar as all other moves had not been notified through proper orders. It took me almost five years to get the orders issued for various moves in spite of repeated rejection of my requests at various levels. Action was finally taken when I threatened to appeal to the highest authority. In the process—**I learnt that perseverance pays and never allow your rights to be curbed.**

The regiment (18 Cavalry) had a glorious past including its performance in the Second World War. It had Sherman III tanks of Second World War vintage and the troops comprised of Rajputs, Jats and Mohammedans, in the three squadrons. The troops were highly trained and full of fighting spirit. At the time of Independence in 1947 and partition, a major portion of the regiment had moved to Pakistan.

The equipment of the regiment was in a state of disrepair, through lack of technical support and maintenance. Most of the senior officers of the regiment including the Commanding Officer Lt Col Balwant Singh had been posted from other armoured regiments. They came from different regimental cultures, with many strengths but plenty of interpersonal problems and conflicts. The redeeming feature was that all the young officers were exceedingly bright, ambitious and full of life. I discovered later that the Commanding Officer was an outstanding and upright soldier. Though a hard task master, he was not popular with some officers who propagated misinformation about him. He had the ability to say 'No' and did not believe in compromises, a true leader with courage and character.

Standing for My Rights

The regiment was asked to move from Jalandar to a location on Jammu-Sialkot (in Pakistan) Road, for operational duties with the 26 Infantry Division. The tanks were moved by train to Pathankot and from there by tank transporters to the location outside Jammu. All soft vehicles (jeeps and trucks) of the regiment and LAD had to be moved by road and a commander for this move was to be appointed. The adjutant of the regiment was being considered, which I resisted as he was junior to me. Finally I commanded the move from Jalandar to Jammu and all officers for the regiment (captain and below) moved under my command. This created the desired impact on the morale of LAD personnel, as never in the past had an LAD officer been given this position. This also gave a message to the officers in the regiment that they could not ride roughshod any more.

As a LAD officer, one was under the administrative control of the commanding officer of the regiment but functionally one had to get support from the commander Electrical and Mechanical Engineers, who was part of the divisional headquarters. The instructions from functional commander (EME) sometimes imposed constraints to meet the operational efficiency of the equipment by creating too many restrictions for indenting spares and undertaking certain types of repairs. It was impossible to meet the objective of ensuring battle worthiness of the regimental equipment, particularly armoured fighting vehicles if one followed all the rules and regulations which had been laid down by EME and ordnance channels. Therefore, personal risks had to be taken to bypass the regulations to meet the main objective. At times one came in conflict with the functional commander in this matrix organisational structure. Once Commander EME 26 Infantry Division called me to divisional headquarters, as he had been informed by CEME First Armoured Division (the formation where 18 Cavalry was located earlier) about the poor maintenance of equipment by 18 Cavalry and ordered me to convey his unhappiness to the commander 18 Cavalry, who was my administrative boss. (I was an integral part of the 18 Cavalry to provide maintenance and repair cover during peace and war.) I smiled and requested him to put himself in my position and appreciate the consequences of his orders to me. He was a bright officer and understood my point. I requested him to attach Armoured Fighting Vehicles increment to our LAD (for additional repair facilities) and assured him of excellent results. This was accepted. In the past, the AFV increment was part of 58 Infantry Workshop Company which did not have officers trained on armoured fighting vehicle repairs. Therefore, armoured regiment did not get satisfactory repair cover.

Fighting Back Petty Politics

In the field location, troops had to live in tents. Condition of these tents was poor. We managed to bring bricks from villages which had been abandoned in the forward areas to build semi-permanent accommodation with our own resources for the troops. This helped to further improve the morale of the troops. This no doubt created a fair amount of heart burning in the squadrons who had failed to use the initiative. A word was passed to the commandant that the LAD was busier in building barracks instead of repairing equipment. One day the commandant appeared in the LAD without notice, accompanied by the regimental risaldar major and asked for me. I had just returned after testing a tank and the commandant saw my condition in working overhauls full of dust. He admonished the risaldar major in my presence, and praised LAD for looking after the welfare of troops apart from the excellent repair cover for the regiment.

Action on the sports and cultural activities enabled LAD to win two trophies in basket and volleyball, beating all the regimental squadrons. This was a very commendable performance which raised the morale of the troops further.

I ensured necessary liaison with other workshops in the Division and developed a good communication with Commander EME 26 Infantry Division and got full technical support. By following unconventional methods, hundred per cent equipment including armoured fighting vehicles (tanks) were brought to operational efficiency, a rare distinction for the 18 Cavalry. In exercise 'Katar', the 18 Cavalry was the only unit in the 15 Corps which had every piece of equipment in excellent shape. The Commanding Officer (Lt Col R.L. Chopra who by then had replaced Lt Col Balwant Singh) received the accolade. I had to take a lot of professional and personal risks to demand spares, totally in contravention of laid down procedures. The motivation was to keep every equipment fit for war.

All officers have to pass Part 'C' Promotion Exam, which is a practical test on battle tactics and is conducted by Infantry/Armour senior officers. As luck would have it, I was the only officer from the 18 Cavalry who passed in the first attempt whereas two majors from the regiment who appeared with me could not qualify. Col Balwant Singh watched the qualitative improvement in every aspect of the performance of LAD. He treated me almost like a younger brother and gave full administrative support to the LAD.

Senior Role Model

One day, Major Hari Singh and myself, while returning after a game of squash on my private motorcycle met with an accident with a bullock cart. Col Balwant Singh came rushing in his private car, not even waiting for the official jeep and picked me up and took me to his house. His wife forced me to take a glass of hot milk to revive myself. There was never a demand from this officer for anything. His generosity in dealing with LAD (EME) personnel at par with the regimental armour personnel was unequalled by anyone else. He was a true leader. His tactical and professional skills were outstanding. He inspired confidence in the troops.

Brig. Tanner (Director EME, a British Officer on deputation) from Army Headquarters, Delhi visited 26 Infantry Division and came for inspection of equipment maintained by the LAD. Lt. Col. Docca CEME (my functional boss) accompanied him. One could see from his body language that he had been put under a lot of pressure by Brig. Tanner, an officer of very high calibre. Lt Col Docca was also a brilliant officer but the inspection of other workshops had not gone off well.

Brig. Tanner was an expert on Armoured Fighting Vehicles, and asked many searching relevant questions. He asked me about our problems which he noted down in his own hand. One could see the smile on Col Docca's face because ours was the last unit being inspected. Col Balwant Singh also gave his appreciation of the EME boys in glowing terms. Col Docca, CEME informed me that Brig. Tanner DEME had identified me for training in the UK on the latest Armoured Fighting Vehicles.

Working in a New Regimental Culture

Col Balwant Singh was posted out. The command was taken over by Col R. L. Chopra, a highly intelligent officer, but very academic and soft in his dealing. He got influenced easily by parochial armour officers. One could see a sea change in the attitude towards EME personnel. As anticipated, relation between the regiment and EME (LAD) deteriorated and regimental equipment got downgraded resulting in serious problems within a few months of my leaving the regiment.

Col Chopra and his wife, mixed a lot socially and had maintained a very fine regimental-friendly culture.

The President of India, Late Dr Rajendra Prasad was to visit the 18 Cavalry. The programme as prepared did not include a visit to LAD. I was told that the president was visiting the armoured regiment. This gave a feeling that LAD was being considered as a separate entity, whereas we had totally identified ourselves with the regiment. I had to provide the necessary inputs to the corps commander about the state of health of the president when he and GOC 26 Division had visited the regiment to finalise the programme. It was decided that the president should not be made to go around the regiment and instead equipment should be demonstrated in the LAD location where plenty of open area was available for the purpose.

The president was received by the commandant, OC LAD (myself) and risaldar major of the regiment. He was shown around the equipment by me. Immediately after the visit of the president, I was deputed to the UK for attachment with Royal EME and BOAR (British Army of Rhine) in Germany. Maj. Hafiz, one of the officers of the regiment who had a tremendous sense of humour, asked me whether I had whispered something in the ears of the president, during his visit.

Dare to Risk

During the movement of Armoured Fighting Vehicles through a river bed, two tanks of the regiment had got submerged in water. The regiment officers tried their best to recover the tanks but failed. They were very scared of the consequences as an enquiry was mandatory for such an accident. They had no option but to come finally to me for help. I along with my boys went and recovered the tanks. We had to literally overhaul the tanks to avoid serious embarrassment and problems for the regiment. No one in the higher command even got to know of the accident.

During recovery of the tanks, my boys had laid the rope way network to winch out the tanks. One of the young bold regimental armoured corps officers wanted to go via the rope way to the tank. It was a dangerous request as he was heavily built. I was not sure about his swimming capabilities either but agreed on his persistence. I however made sure to position two of my best swimmers downstream. As anticipated he dropped like a brick midway. All of us missed a heart beat. However it got converted into great laughter and jubilation by a large group of regimental officers and soldiers from the LAD and regiment, when the officer was rescued by my swimmers and brought out of water.

Relationship Management

I had an opportunity to play cycle polo with armour officers. I asked Major Hafiz one day as to why cavalry officers had come down to cycle polo from horse polo. He replied, "Nowadays when we throw a ball in front of the horse, the horse wants to eat it. Where is the money to look after the horses". I had made some lasting friendships with the Cavalry officers. There was a galaxy of them. Each one had strengths, but two stood out—2nd Lt Kailash Dhody (retired as major general), who had just joined the regiment and Lt Gurdev Singh Kler (retired as lieutenant general). The latter was very keen professionally and had excellent human relation skills and a sense of humour. Kailash impressed me with a sound and refined background. He had excellent command over many languages and was well versed in music and poetry. Kailash wrote a speech for me in Urdu when I was asked to be a master of ceremony in an EME function in 26 Infantry Division. He had thrown in some Urdu couplets, which I used with some effect in my speech and later in life as well.

This tenure of over two years with the regiment gave me tremendous insight into the minds of different people. Depending on the conditions and movements on the Pakistan side, the regiment was often put on six hours notice for move to operational area. I also had the unique experience of having to deal almost independently with two bosses in the field, one dealing with operation matters while the other monitored the technical efficiency.

3

On Foreign Deputation

I left for London by sea (P&O Ship CORFU) on 18 May 1956 for training with REME in UK and BOAR (British Army of Rhine) in Germany. In UK, I was attached to REME Training Battalion BORDON, and had attachments with two Army base workshops at Aldershot and Chilwell (Nottingham), Royal Ordnance Factory Barnbow, David Brown Gear Box, Rover Motors and a field workshop in Germany. It was a great experience in maintenance and repair of armoured fighting vehicles. Having worked with British army officers and technicians of all ranks, I found very little difference in capabilities except that they had easy access to technology being in a highly developed country.

In Nottingham (for training with REME base workshop), I stayed in George Hotel, managed by a gentleman and a lady (not related to each other), both with highly refined personalities. The gentleman was a bachelor with a hobby of collecting antiques. Each room of the hotel had a different ambience. I stayed there for almost eight weeks and got very friendly with the management, who made me very comfortable beyond the formal contract.

It was a thrilling experience to be served by English staff during the voyage, stay in England, in army officers' messes and while travelling for sightseeing in UK. This was my first visit abroad. I developed quite a few friendships which were useful during subsequent visits. Some of the REME officers who had worked in India were very generous in their hospitality. They would even come and drive me to their homes and after meals drop me back. In Germany, I had very close interaction with the REME officers and their families. I enjoyed their hospitality and found them very warm after the initial reservations were removed through their perception of my professional ability to take them on equal terms. A rare compliment was given when the young officers of the REME workshop bid me farewell by giving a formal salute.

At Aldershot base workshop, I stayed in the REME officers' mess along with two Pakistani and two Iraqi officers. The English batman used to get confused with Indians not eating beef and Pakistanis not eating pork. We were made to sit on opposite sides of the dining table to avoid mistakes in serving of meat. One day stew had been served and while we were half way through the meal the batman came running and asked us to stop eating as by mistake he had reversed the dishes—we had no option but to smile!

During the assembly of a gear box with aluminium casing, one of us had over tightened one of the studs which got stripped from the casing. It had over sixty studs of this type. The *Iraqi* officer continued to assemble the casing but the British instructor asked him to strip the gear box and said, "Sir, I will not allow this gear box on any tank to endanger the life of my son who may be on this tank during war." A very strong lesson on quality was emphasised.

There was a requirement for getting drawings of jigs and fixtures for use for the maintenance and overhaul of Armoured Fighting Vehicles in India. The formal procedure of obtaining these would have been time-consuming. Good social contact with REME officers and their families did help.

I returned to India on 18 April 1957 by sea via the Cape, as the Suez Canal was closed. I travelled by Cilica Anchor lines. The journey took almost a month. The passengers were from different nationalities and were a healthy mix of young and old, of all ages and genders. We enjoyed playing games in the day and dancing at night. On the way, I got down both at Durban and Karachi. I had asked the taxi driver at Karachi about the conditions in Pakistan. He said in Urdu, "Ever since the death of Liaquat Ali, Rishwat Ali (corruption) has come in. Earlier we were slaves now we are slaves of slaves". The situation in India seems to be no different even today.

4

The Instructional Years

On my return to India, I was posted as an Instructor in the EME School at Trimulgery, Secunderabad. After returning from UK, I got married on 12 May 1957 to Shobhana. I proceeded to Secunderabad with my wife and mother, and stayed at a private accommodation there. I was allotted a government accommodation on the top of a hillock on one side of which was a cemetery and on the other side a burial ground. It was a lonely house which I refused to accept. I was marched before the Dy Commandant Col Kartar Singh who with all seriousness said, "We have given you an exclusive house, and you have refused to accept". I said, "Sir, I have not refused but have asked for a guard of one NCO and three Sentries at day time and two and six at night, as I am on camps most of the time training student officers, and will have to leave two ladies in this desolate place." He smiled and said, "I understand". A beautiful bungalow fell vacant thereafter and I requested for allotment. Commandant Brig. Rao sent my name to the headquarter's sub area but strongly recommended the name of his adjutant who was junior to me. The bungalow was allotted to me, as my plight was known to the subarea. Brig. Rao was under the impression that I had pulled some strings. I was the first in the family to join the Army and did not have any 'uncles' in the hierarchy. Brig. Rao was very unhappy and told me in front of a couple of officers, "I do not like officers who go from the back door". I replied in the same sarcastic tone "Sir, what does one do when the front door is closed". He walked away with a frown. I have never bowed down to anyone and have learnt to stand on my own legs and fight my own battles.

I was initially in charge of recovery and waterproofing training, and later on was responsible for training on technology, repairs and maintenance of Armoured Fighting Vehicles. There was a reorganisation in the corps of EME and a number of additional vacancies of majors got created. There were eight captains in the EME School including me who were due for promotion. The management worked out a strategy to promote seven, and I had to await an enhanced organisational structure for the EME School. I was furious, obviously as officers junior to me by one year would have been promoted, while I would have been awaiting for the upgrading of EME School which was nowhere in sight. I therefore decided to put in my resignation letter and proceeded to the EME School Officers' Mess, where a lunch had been organised for the Deputy DEME from Delhi. Within the hearing of many senior officers, I deliberately talked about the lack of concern for the morale and motivation of the junior officers.

> I have never bowed down to anyone and have learnt to stand on my own legs and fight my own battles.

I was marched before the DEME Brig. S.P. Vohra next morning. I was again asked about my problem to which I replied, "Sir I am considered to be an indispensable officer, but will be watching with frustration officers who are one year junior to me getting promoted. There is an officer who has just returned from UK after completing a course similar to mine. He can very easily replace me." Commandant Brig. Rao who was present had no answer. I was posted to Delhi in EME Directorate Army Headquarters on promotion in Vehicle Technical Policy Group.

At the time of Independence the seniormost Indian officers in EME were comparatively very junior as British officers had returned to the UK. When last of the British DEME Brig. Tanner left, Maj. Gen. Harkirat Singh from the corps of engineers was posted as director EME. He wanted to familiarise himself and worked out a 10 days' programme to visit Poona, Bombay and Jabalpur. I was asked to accompany him as his staff officer.

I did not have much exposure to the corps. I was now being sent with the seniormost officers of the corps. I was too young to understand the motive at that time. Looking back, the idea was to fix me as no one wanted to go near the general for fear of being exposed. For me, as it proved later, was a blessing in disguise as I learned a lot by watching at close quarters the style of working of this brilliant officer. Apart from being brilliant, he was generous and made me join in all the social activities which were organised in his honour. I used to smoke but was no match for this Sikh officer smoking away State Express 555 cigarettes. I had the honour later, on 7 April 1994, to deliver the 10th General Harkirat Memorial lecture organised by the Institution of Engineers, Delhi.

Infantry School Training

Through job rotation and variety of training programmes in different functional areas officers are trained to shoulder higher command leadership positions in peace and war. I was detailed to attend Junior Commanders' Course, an all arms/services training programme at Mhow. Col Sucha Singh, Chief Instructor, a highly decorated and experienced infantry officer was impressed with my performance. I was given command of a platoon in a night attack exercise, a rare distinction for an EME officer, much to the chagrin of infantry and other fighting arm officers.

Bureaucratic Working

Gen. K. S. Thimmaya, as the chief of army staff, during a talk to the officers stationed at Secunderabad/Hyderabad had said, "Gentlemen all this talk of Hindi-Chinee Bhai Bhai is 'Bunkum'. Our real threat is from the north". This was some time in 1958/59. The army had to prepare for operation of equipment and vehicles at high altitudes. Late Brig. MML Chabbra (retired as Lt Gen. DEME) who was DDEME asked me to prepare two papers, one on requirement of super chargers for high altitude operations for vehicles and the other for requirement of oxygen for troops at high altitude. These were required for getting financial sanction for import of equipment for trials through the famous hurdles of bureaucracy of the government.

For the first paper I had the technical background but the second was a pure medical subject. I was advised during my earlier stay in the 19 Division by one Major Mahatab Singh from state forces, who had many anecdotes from his experience with rajas and maharajas, that in the army one should never volunteer, but also never refuse when some task/ responsibility is given. It is a golden principle, but as a leader of a group if one volunteers to lead, the group will follow irrespective of the dangers involved.

I had to undertake basic research on these subjects. I visited many libraries and finally met Mr Dhanpat Rai at the Defence

Research Laboratory library in Metcalfe House, Delhi. I have still to come across a more knowledgeable and committed librarian. He produced literature which was relevant and useful on both the subjects. My task was made easy. After preparation of these papers, I was considered an authority on the subjects, and was invited by many senior officers for discussions.

Brig. M.M.L Chabbra asked me to meet the scientific advisor to the Defence ministry, to provide any clarification on the file for import of super chargers for the operation of automobiles at high altitudes. I was explained repeatedly that the scientific advisor was a very senior officer. I had to be careful in talking and not shoot from the hip. My reputation had travelled all over!

I went at the appointed time and met the gentleman. After listening to me he asked if he could call some of his junior colleagues. I had to almost give a lecture to four scientists starting from the fundamentals. The file got cleared and I got my first direct exposure to the bureaucratic working and the growing frustrations of those committed to performance targets.

My direct boss was Lt Col K.S.K. Murthy, with whom I had a brush in armoured division due to my brash behaviour. I found him exceedingly brilliant, generous and a kind person. On his transfer another bright colonel was appointed in his place. He was already familiar with the bureaucratic working in which he also believed. He would clear those files where directions of higher authorities were required. The rest of the files he would send back with remarks – 'Please speak/please discuss', but he would have no time to speak or discuss! I bundled all these files and put them on his table and pinned a note 'Please call as and when you have time'. He fired his personal assistant for allowing the files to be kept on his table. The personal assistant told him that Maj. Wahi himself had brought the files. If a peon had brought then he could have stopped him. He asked his personal assistant to distribute the files to those concerned and asked them to bring the files directly to him. No files were getting cleared. One assistant by the name Mast Ram came to me and said he had taken the file to the colonel 10 times, but there was no decision. I told him to put up a note on collation cover file with the dates he had gone to the colonel for directions. I put my own note on the file and mentioned that this was one of the many files on which reminders were being received from other departments.

He sent for me and said, "You think I do not know my job". I replied, "Sir, you know your job very well, but we need your advice. In case you feel that no file should come up to you, then I shall clear the files." He agreed to meet me one hour before time every morning. I went to his office next morning as directed but he started reading every page of the file; not one file was cleared! In the meantime his posting order was received and he wanted to avail two months leave. I requested him to write confidential reports (CRs) of all the officers before he proceeded on leave. He asked what was the hurry. I was the seniormost major under him. So I told him that it was the desire of all officers that their CRs should not be delayed. So he asked me to inform other majors to meet him at 5 p.m. He gave me an outstanding report but for another major who used to walk around his office he remarked, "This officer does not know how to assert himself. He was a very good officer but always under tension because of the behaviour of the boss."

5

Deputation to Heavy Vehicles Factory —Avadi

We heard that a heavy vehicles factory was being set up at Avadi for the manufacture of armoured fighting vehicles. Officers on deputation from EME were requested by the department of defence production. I also volunteered to go on deputation along with a few others. The Director EME (Maj. Gen. S. P. Vohra) used to have an informal meeting with officers over tea every Saturday. In one of those meetings, he told me that he could not afford to spare me on deputation to the heavy vehicles factory. I was however assured that I would continue to officiate as Lt Col till I was in the army headquarters and enjoy additional officiating pay.

After a few months, Gen. Vohra called me and said that he had agreed to release me on deputation to the heavy vehicles factory. Mr Rao who was working in Technical Development (TD 25) and used to interact with me on technical policy matters met me and told me the real story behind my deputation. It was no act of generosity by EME, but Mr Mantosh Sondhi, head of the project, had checked with his colleagues from Technical Development Directorate about the suitability of EME officers. My name was suggested by almost everyone. Mr Sondhi had asked Mr Rao to tell me to meet him since the posting order on deputation was under issue. This was the first time I met him. I had explained to him that I would be losing officiating allowance, which I was drawing for working in the vacancy of a lieutenant colonel. He just smiled. I was deeply impressed with his mature interaction with me.

Mantosh Sondhi had the experience of working with service officers, particularly army officers. He was a great admirer of their professional and leadership qualities, in particular their ability to effectively manage human resource. I had a life-long association with him until his death and consider him as one of my main mentors in my career.

Heavy Vehicles Factory at Avadi

The factory was in the stage of planning and construction when I had joined in November 1962. I was involved with the procurement of machine tools through DGS & D (government organisation for centralised high value purchases) for the manufacture of components and assemblies such as engine, gear box and suspensions other than fabrication and machining of hull and turret. I had to interact with the collaborators, Vickers Armstrong, Leyland Motors and Self Change Gears in UK, to obtain technical details of machining operations, requirement of tools etc. It was a colossal task but was achieved according to time schedule with minimum manpower. The DGS & D was known to be a very bureaucratic and slow moving organisation. A very close liaison, including literally doing their basic work of preparing comparative statements of offers received against each tender, had to be done to stick to the time schedule.

After the completion of work on procurement at Delhi, I moved to Avadi to ensure receipt and installation of machine tools. Initially I had an improvised office by using packing cases of large machine tools on the shop floor—a strategy which ensured supervision on the spot to ensure commissioning of machine tools as per schedule. The machines had been procured from UK, France, Switzerland, US, Germany and Japan. I had to visit works of the suppliers to ensure proper coordination for jigs/fixtures and tools and receipt of equipment in time. I was also deputed to the works of the three collaborators (Vickers Armstrong, Leyland Motors and Self Change Gears) for familiarisation with the equipment and training for production.

A very modern factory for the manufacture of Vijayanta tanks was established covering fabrication/machining of hulls and turrets, manufacture of L60 engines, epicyclic gear box, a modern forge shop, tool room and engine testing facilities. One worked on the shop floor with the workers to ensure quality and time schedules to produce the first indigenous tank. It was a great experience in production technology, human resource management and general management.

For the first time in the country an armoured fighting vehicle (tank) was manufactured. There were stream of visitors of all levels from the services and government. Invariably I had to conduct the visitors along with the general manager.

The workers were recruited from all over the country and trained in the artisan training school set-up at Avadi. Some were also trained at the works of the collaborators' machine tool suppliers. Most of the workers and supervisors were young but highly motivated. They ensured excellent results. The manufacture of Vijayanta tank was on schedule, within budget and against all odds. The factory was inaugurated by Indira Gandhi on 3 July 1966. I had the honour of escorting her from the Madras (now Chennai) airport to the factory at Avadi.

Organisational Culture

The credit for excellent performance goes to Padma Shri Mr Mantosh Sondhi, the general manager, an outstanding technocrat who had a unique style of motivating people and following up their activities meticulously without breathing down their necks. He also had an excellent way of managing people in the government and associated departments, including DGS & D to avoid road blocks to the progress. He was a strategist par excellence.

He interacted at all levels and listened to the problems including personal ones. He went out of his way to solve even the personal problems of people. He made sure that I got my promotion to the rank of lieutenant colonel, even though I had not done the criteria appointment of commanding a workshop as a major. **He was an intellectual giant and a good human being with unimpeachable character and integrity, courage, commitment and concern for the organisation and the people.**

Mr Sondhi had handpicked officers from the government, army, navy and ordnance factories and integrated them into an excellent team. Each member put in his best according to his potential. He was a role model as a leader and set a personal example. He also developed leaders under him. He won the genuine respect of his junior colleagues and workers. **His wife Mrs Rita Sondhi was a source of inspiration through her charm and she took great interest in the welfare of the families of the staff and workers.**

She made significant contribution to create the desired environment for people to put in their best.

We had excellent social life at Avadi. We were members of the Madras Gymkhana, and used to go there every Sunday to play Bridge. My son and daughter enjoyed swimming. We used to enjoy Dosas in Woodlands regularly. Our youngest child Shalini was born on 20 February 1968. A few days later I left for Ambala to do my criteria appointments of commanding an EME Battalion.

6

Back to the Army

The 6014 EME Battalion was being raised as part of 14 Independent Armoured Brigade, called Black Charger. It was the fastest mobile hard hitting formation to be used to break through in attack or defence in depth. It had three regiments, 9 Deccan Horse, 64 Cavalry and 18 Cavalry with which I had the honour of serving earlier. The brigade commander was Brigadier R.P.S. Randhawa, who had earlier been in the 18 Cavalry. We had not worked together but had known of each other's reputation. He was one of the finest leaders and regal in his behaviour. He was outstanding professionally and a tough task master.

Effective Leadership

The 6014 EME Battalion was under raising. Officers, JCOs and NCOs were being posted from other units. We had a mixed manpower, as no unit was keen to spare their capable men in a hurry. I believed and still believe that people perform and put in their best in proper work environment of an organisation, which is created by proper leadership. Everyone has to be treated as a gentleman and an outstanding performer till by their results and behaviour they come down from expectations. Two of the senior majors had been posted under me with average performance reports. They had shown total commitment and produced outstanding results in view of the participative culture which was established. Both these officers later reached the level of general officers. The morale of the battalion was raised by close interaction and by taking personal interest in the welfare of the troops. Sports activities were given a big thrust. During peace time the morale of the battalions/regiments is judged by their performance in the sports. Our battalion won three trophies (hockey, volleyball and basketball) during the annual sports in the brigade.

The brigade had to move out of its location for an exercise. The brigade commander asked my battalion to remain in location to complete its raising. One early morning, when I was witnessing the PT parade of the battalion, the staff officer of the brigade commander came and conveyed the desire of the brigade commander to meet me at the exercise area immediately. The brigade commander mentioned that some of the tanks (T54 Russian) had developed a defect in the connecting gear box. He wanted me to inspect and advice. On inspection I found that it was a design defect. All tanks would suffer from this defect in due course.

The brigade was operationally not fit for war. The normal procedure would have been to make a defect report and send to higher technical authorities. The Technical Development Establishment at Ahmednagar would then investigate the defect and take perhaps months to suggest a modification. Repair by base workshops would then take about a year and the whole formation would remain unfit for war.

The brigade commander was obviously worried and asked me whether I could handle it. I told him that my battalion boys would do the job if we had his support as I would be bypassing all laid down procedures. I discussed with my officers and decided on the modification. The brigade commander wanted that tanks should be modified in the exercise area so that troops do not lose faith in the equipment. A vital factor in the Army is to ensure the high morale of the troops. The brigade commander had given me full powers over his staff to get support to complete the task. Maj. Manmohan Singh (retired as Lt.Gen.), Maj. Nijhawan (retired as Maj.Gen.), Maj. M.L. Khanna (retired as Maj. Gen.) and other officers, JCOs and craftsmen in working overalls were on the job round the clock to complete the task. We completed the modification in 10 days' time. In the meantime, the defect report was raised. A technical team from TDEV-Technical Development Establishment Vehicles, Ahmednagar arrived and confirmed the modification completed by us. The corps commander, legendary Lt. Gen. P. S. Bhagat VC also visited the brigade in the exercise area and appreciated the excellent work done by my battalion.

In a rare gesture, Brig. R.P.S. Randhawa recognised my contribution by asking me to give away prizes in the brigade's boxing competition organised in 9 Deccan Horse. The brigade commander took leave on that date to ensure my giving away the prizes. The whole brigade stood up after the prize distribution to say, 'Col. Wahi Ki Jai'–a rare honour for an EME officer. Extracts from two D.O. letters appreciating the work done are appended below.

My dear SP 22 May 1968

I am writing to you with regard to the modification carried out to the Connecting-Gear Assembly of the T-54 Tanks of the Brigade. I have received a DO from GOC XI Corps an extract from which is as follows:-

"I am however, pleased to know that prompt action was taken to inspect all the tanks and an interim modification was thought out. I would like you to convey my appreciation to your CEME, officers and men of 603 EME Battalion for their efforts for taking speedy remedial measures and putting all the affected tanks on road".

It gives me great pleasures to add to it, my own appreciation of the work undertaken and the zeal and thoroughness with which it was completed. I am fully confident that with your guidance, the Battalion will continue to 'rise to the occasion' and ensure the battle worthiness of this Formation at all times. I would like you to convey this to all ranks of your battalion.

My dear SP 29 November 1968

I have received a letter of appreciation from Commander 1 Armoured Brigade for the excellent work that the personnel of your Battalion did during Exercise 'NOVEMBER AGAIN'. He has praised their technical skill and efficiency. Captain C.S. Thomas created a very favourable impression on them all and I have already written to him about this.

I am very pleased with the performance of the personnel of your detachment during this Exercise and would like my appreciation of their work to be conveyed to them. I am fully aware of the enormous drawbacks and handicaps which your unit had to face during the early stages of its raising, specially when I called upon you to take up your full commitments in the Brigade much before your unit was in anyway organized to cope with such assignments. I am rather proud to say that with such tenuous resources as could be mustered, the Battalion has always fulfilled its commitments.

With best wishes,

Sd/- R.P.S. Randhawa

Brig. R.P.S. Randhawa was transferred and was replaced by another brigadier, who was an antithesis in his values, attitudes and competence. Brig. R.P.S. Randhawa, in his DO letter dated 25 April 1969 wrote as follows:

My dear SP

It was very nice to have you with me in Black Charger family. Your responses and contributions were outstanding. I am confident there will be more occasions when we shall work together.

Sd/- R.P.S. Randhawa

Alas! this was not to be. We lost this brilliant officer in a road accident when he was on his way to pick up his rank of major general as commander of an armoured division.

My tenure with 6014 EME battalion came to an end on the receipt of Government orders to move to Bokaro Steel Plant on deputation.

Lessons for Industrial Life

Army training and experience brings out clearly the necessity for everyone to have the same objective—to defeat the enemy. To achieve this objective, everyone has to pull and push in the same direction. There has to be a collaborative approach and no confrontation within the troops. Individual talent and skills in extra curricular activities are appreciated, recognised and rewarded. Acts of bravery and exemplary performance against the enemy are recognised through various awards. Everyone understands the consequences of defeat in battle/war. Leadership of the highest order is essential for success. This is true for success in industrial situation as well, where the objective has to ensure optimum utilisation of resources to generate surplus money for further growth. Any organisation which does not grow will stagnate and perish.

Army to Corporate World

7

Beginning with the Corporate Phase – Bokaro Steel Plant

In November 1969 on a request from the managing director of Bokaro Steel Plant, which was under construction, I was sent on deputation as chief of inspection to ensure receipt of equipment from indigenous sources of the right quality. Sixty per cent of the equipment was ordered on indigenous sources. The plant was being set up with Soviet collaboration. It was initially to be of 1.7 million metric tonne capacity, as an integrated steel plant, with blast furnaces, LD converters for making steel, hot and cold rolling mills, thermal power plant, coke oven batteries, raw material reclaiming plant, heavy machining facilities for repairs, etc.

Organisational Concept

My job was upgraded and I was designated as the chief of planning, progress and inspection. A new concept had emerged by integrating progress and inspection. Officers were transferred to my department from the construction department with different levels of competence, attitudes and behaviour. A lot of effort had to be put in to integrate them into a highly motivated team. As an example, one officer would lock himself in his office in the morning for about one and a half hours, ostensibly for completing his prayers. This officer was an expert on material handling, but was always put on odd jobs. He had completely lost interest in work. A lot of time and effort was put in to bring him back on track. He proved to be an asset!

Mr Mantosh Sondhi was the managing director of Bokaro Steel Plant. **He was a role model in integrity and methodical hard**

work. He had the knack of getting the best out of every individual and managed to seek assistance from the Government, Soviets, contractors and suppliers for the timely completion of the Plant. His relationship management was based on sincerity and strong values. He ensured the construction of the first blast furnace in a record time. The construction of the Plant was behind schedule for over a year when Mr Sondhi was appointed as the managing director. The price of land around the plant escalated manifold within a year of his taking over as the community could now see the early completion of the Plant and the resultant prosperity all around.

Aggressive Equipment Follow-Up

It was a colossal task to get the equipment needed for the construction as per schedule particularly from Heavy Engineering Corporation (HEC), Ranchi. Orders were placed on Heavy Engineering Corporation, a number of other public sector companies and hundreds of private sector companies spread all over the country. I had positioned a group of my inspectors headed by a senior officer at HEC Ranchi to ensure the manufacture of equipment to the desired quality and as per schedule. Heavy Engineering Cooperation had a number of senior officers from the Railways. One day I asked Works Manager Mr Prem Singh, who was from the Railways, "You are all from Railways and still fight with each other". His reply was "We are all from different railways!" He had a good sense of humour and was helpful.

The senior executives of HEC had very little contact with the realities on the shop floor in view of their bureaucratic style of working. We had to get the work done with very close liaison with the foreman on the shop floor, sometimes in fact interacting directly with the machinists, motivating them by providing with the data about the importance of the work being done by them. The workers of HEC had excellent skills and could have been great assets under good leadership.

Relationship Development with Experts

The construction of Bokaro Steel Plant was in full swing. During one of my visits to the plant site, I saw a group of Soviet experts and Indian engineers having very animated discussions on the indifferent measurement results of a 'small bell' for the first blast furnace. This small bell had been received from the Soviet Union. They were getting different results, when measured at two different times. I ventured to tell them that they could never get consistent results by placing so heavy a bell on such soft ground during inspection. The small bell was shifted to a hard surface and consistent measurement results could thereof be obtained. All of them had missed a small but important point. After this incident the Chief Soviet Expert Mr Eroian became a very good friend of mine and started sharing with me several confidential data and information.

Diplomacy in Negotiation

I was deputed to the Soviet Union, along with Joint Secretary Ministry of Steel Dr S.S. Sidhu, (a brilliant IAS officer) to expedite supplies. The Soviets did not believe in giving definite delivery dates for individual items of equipment which resulted in bottlenecks during construction. We had discussions with Tyazhpromexport senior functionaries, headed by Mr Litvinenko (he in the 90s was minister in the Russian embassy in India). We were not making much headway during discussions as the chemistry of Mr. Litvinenko and Dr Sidhu did not match. At the Indian Embassy, an 'at home' session was organised a day prior to our departure. I advised Dr Sidhu to attend the reception and leave me to talk with Mr Litvinenko. We sat for a few hours and hammered out an agreement which covered delivery dates for every item. This was a departure by the Soviets from their normal practice, much to the satisfaction of all concerned from the Indian side.

Mr Eroian, the head of Soviet team with Bokaro Steel Ltd, who was back in Moscow after completion of his tenure in India, was also present in one of the meetings. He came in and sat with the Indian team, next to me and supported every argument put forward by us. Next day, he was nowhere to be seen. These were the ways of Soviet Union authorities.

Surveillance by Soviets on Sensitive Visitors

In one of the visits, I was leading a delegation of public sector executives to Soviet Union. We had to visit Leningard. One Mr Mushkin was deputed as our guide. He was about 6' 3" tall, very lean and had a serious poker face. To every question, his reply would be 'may be possible'. While shopping for toys, we could see a gleam in his eyes. On enquiry we discovered he loved children. We bought a few toys and gifted them to him. He accepted them very reluctantly. I asked him to go to the hotel and leave all the purchase. He agreed to do so after he had made me and my colleague promise to stay put in two chairs in a restaurant on the ground floor, till he returned.

After his departure, we shifted to a new location on the first floor to see his reactions on return. We saw him returning and going into the restaurant and rushing out in bullet speed after finding us missing. He was running up and down the road like a mad man. After some time we shouted for him. The moment he saw us, we saw the grin on his face. We asked him to keep smiling. He replied, 'May be possible'. Later on we discovered that he belonged to the Soviet secret service and was always made to accompany sensitive visitors who could collect data beyond what was allowed by the authorities.

Blast Furnace Commissioned on Schedule

The first blast furnace was commissioned as per schedule and inaugurated by Indira Gandhi. Thereafter, Mr Mantosh Sondhi moved as Secretary Heavy Industries, Government of India, Delhi. Dr Bhattacharya an eminent metallurgist was appointed Managing Director Bokaro Steel Ltd. Mr K.C. Khanna took over as General Manager (Operations) and Mr S. Kumar took over as General Manager (Constructions). I was the Chief of Planning, Progress and Inspection. We three (along with Mr Khanna and Mr Kumar) were at the same level and all four from the same university (BHU). The style and tempo of working had completely changed. A bureaucratic culture had set in, as happens in many large organisations. Mr K.C. Khanna was the seniormost amongst the three of us and was appointed as managing director after the retirement of Dr Bhattacharya. I was therefore looking out for a change. I received a letter from Mr Wadud Khan Chairman SAIL and Secretary (Steel) in the Government who agreed to my new assignment at BHEL.

My dear Wahi,

Your letter of 6th April is greatly appreciated. About releasing you, I had taken the decision in response to the request you had made through Khanna to be relieved from the Bokaro Steel Ltd., to take up a better and independent position in the public sector. I mentioned to Khanna that I was extremely reluctant to let you go but I changed my views on being advised that you were moving to an independent charge. I also learnt that location-wise the new assignment suited you more personally.

Although we have not met on more than a few occasions, from the reports I have received and from my personal observation, I formed the clear opinion that you were a valuable asset in Bokaro, and for that matter in any industrial enterprise.

I wish you every luck and happiness in your new assignment. I have no doubt, you will distinguish yourself in providing the right stewardship to the affairs and work in BHEL.

With all good wishes and kind regards,

Sd/- M.A. Wadud Khan
April 16, 1974

8

Leading BHEL (Haridwar) to New Heights

After my tenure at Bokaro, I was picked up by Mr V. Krishnamurty CMD BHEL, to set up the Central Foundry Forge Plant (CFFP) a unit of BHEL at Haridwar, as project administrator and was later designated as executive director. I had taken up the assignment in March 1974. This plant was conceived earlier and a project report had been prepared, but construction activities could not start due to bureaucratic delays. M/s Crusot Loire of France were the collaborators.

Mr Krishnamurty had the vision to make BHEL grow to the best international standards. He, therefore, revived the CFFP project and visited France to re-initiate the interest of the collaborators. I, along with a few of my colleagues, had preceded him to brief the collaborators about our intention and will to complete the project. Crusot Loire was a large conglomerate of a number of factories at different places producing high grade steel and equipment. They also had major interest in Framatome, a group manufacturing nuclear equipment. A very close liaison was maintained with the collaborators, by deputing executives and shop floor personnel for training. I, along with other senior colleagues, also visited them periodically. The collaborators cooperated well. Excellent living accommodation and other facilities were provided at BHEL Haridwar for them and representatives of firms supplying equipment and services. This enabled us to get their commitment to meet our objectives by getting them emotionally involved with BHEL.

Heavy Electrical Equipment Plant of BHEL (HEEP) was involved in the manufacture of thermal turbines and generators up to 235 MW, hydro generators and turbines of various sizes, AC/DC motors and control equipment. This plant was established in collaboration with the Soviet Union in 1960s and was headed by a general manager, designated as executive director later. The Central Foundry Forge Plant was set up next to HEEP.

New Organisation Concept during Construction

To complete CFFP construction some executives were transferred from other units of BHEL and the balance were recruited from the market. It was decided that a separate construction department will not be created. To bring in ownership concept right from the beginning, the future operational managers would act as construction managers of their facilities. This strategy was conceived to avoid problems of handing over from construction to operation departments and to avoid surplus manpower after the construction phase, which normally resulted in labour problems.

This also ensured total commitment of managers to quality and completion of facilities in time, as per the requirement of the operational managers themselves. Through this system, operational managers had the freedom to make changes during construction. This enabled avoidance of conflicts between construction and operation people, which takes place when a separate construction department is created during project construction stage. This also brought in ownership concept for the operational managers, which is so essential for quick decision-making and accountability.

Human Resource Management

The workers were also assigned as artisans and not given any specific designation to ensure job enrichment and ability to shift artisans from one job to another without labour problems. Everyone from the highest to the lowest ranks was on the factory site to ensure quick decision making through constant interaction. A lot of outsourcing was done and contractors/suppliers were given due respect and treated as part of the system. This enabled everyone to own the targets for completion. The staff and workers from Jessops Calcutta who were mostly from Bengal worked even on Puja days to complete erection of a crane which was vital for the commissioning of the plant. This was a rare sacrifice which would not have been possible through any monetary incentives except through emotional involvement by giving them a sense of belonging.

Total transparency was ensured in equipment/material purchase. In this context, I narrate here one incident. GEC and Philips had quoted for lighting equipment. A representation was received after placement of an order. It was noted that due to an error the order was not placed on the deserving party. After a presentation was made by a young executive, the order was rectified. This very young executive met me in Dubai after over 27 years. His letter, as reproduced below, had touched my heart immensely.

Dear Col.　　　　　　　　　　　　　January 2003, Dubai (UAE)

Was a delight meeting you again. Twenty-seven years is a long time — but through these years I have always seen you as a beacon of light in a country shrouded in darkness. You set by example, the highest standards of integrity and left a mark on an impressionable young executive myself.

God bless and help you and give you peace—Amen.

With warmest regards,

　　　　　　　　　　　　　　　　　　　　　Sd/Steven Pinto

Employees' Commitment

The sense of discipline exhibited by the senior executives was exemplary. One day, I, along with my general manager Mr Rao was on a round of the project site. We saw a trailer with very heavy load being pushed through an obstacle by about 30 workers. They were making a lot of noise for pushing, but the trailer did not move an inch. Mr Rao and I went and just put our hands on it. The trailer moved out of the obstacle with one push. Ours was only a symbolic effort, but it had an electrifying effect. Once, senior executives identify themselves with work, no matter how small it may look, it raises the morale and motivation of the workers. It brings out respect and dignity for physical work.

On another occasion, when I had finished talking to a very young executive, Mr Rao, the general manager who was also with me, commented on the lack of discipline of the young executive because he was standing in a very casual manner. Mr Rao had a very high standard of conservative values. This not only brought out the necessity for training of young executives on attitudes and behaviour, but also brought out the need for seniors, to have with time more tolerance for behavioural changes.

A very open system of management was established and senior executives were easily accessible for discussions and quick decision-making. Executives from different departments sat together for decision-making to avoid floating of files between departments.

> ❗ Once, senior executives identify themselves with work...it raises the morale and motivation of the workers. ❗

Project Completed in Record Time

The project was completed in a world record time of 16 months to the pleasant surprise of our collaborators and the BHEL Board. CMD BHEL Dr Krishnamurty had inaugurated the plant along with full board members of BHEL and Mr Ferry. It was a highly sophisticated plant where creep resistant steel for turbine parts and rotors had to be produced through state of art technology and equipment. Excellent test laboratories and detailed quality control system had to be established.

Thereafter, castings and forgings of international standard were produced and we exported casings to Westinghouse of America within six months of the start of production. They agreed to give 20 per cent increase in prices for a repeat order on their own. This proved beyond doubt the capabilities of Indian workers and managers to achieve international benchmarks, given the desired work environment created by the leadership.

Mishra Dhatu Nigam Hyderabad was being established, also with the collaboration of M/s Creusot Loire, our collaborators for CFFP–BHEL. The project was not moving at the desired pace. I was inducted in the board with a view to share my experience of project management.

Integration of Two Work Cultures

On completion of the CFFP project, I was appointed executive director of both the BHEL plants—Heavy Electrical Equipment Plant and Central Foundry Forge Plant, at Haridwar. The very first production meeting in HEEP gave me the feeling that there was total disregard for time. Also the production and engineering departments were at loggerheads. There were a large number of union leaders and their close associates who did not add value to the organisation and caused much disturbance at work.

It was decided that an attitudinal survey needed to be conducted. An elaborate questionnaire prepared by Deputy General Manager, Management Services, Mr G. Saran was sent to over 1,000 executives up to the level of managers. The covering letter of the same as written by the author has been reproduced below.

My dear

1. The KEY ASSETS of any organisation are the men behind it and the organisation effectiveness is really a sum and substance of the thinking of the executives who man it. Its growth or decay could thus be related directly to the degree of positive or negative involvement of the associated group of people.

2. BHEL Haridwar has come a long way and has emerged as a symbol of change for our usually referred to agrarian society. To those involved in this process of change, there are greater challenges ahead. Our role, and particularly the contribution from our younger colleagues has a direct bearing on the results of the experiments which we are conducting in this environment for setting new trends and styles of management.

3. To us, your aspirations and simultaneous fulfilment of organisation needs are equally important factors and we thought nothing could be better than to ask you. With a view to take stock of the situation, I would like to know your free and open views in the enclosed questionnaire. Keeping fears and apprehensions away, please do share your thoughts, personal goals and hurdles as you see them. This would go a long way in our jointly evolving a management style which will be appropriate to our needs and befit our Chairman's thinking who preferred to call BHEL Haridwar as his laboratory for trying out new styles of management, during his recent visit.

With best wishes,

Yours sincerely,
Sd/ S.P. Wahi

The replies received were very revealing. Interpersonal problems were affecting the morale. Quality and training departments were full of executives who were thought to be incompetent. The problems were frankly discussed with more than 500 executives, who had very negative/positive responses. For instance, a young officer mentioned that his boss was always critical. I pointed out that it was his boss however, who had recommended his name for training abroad. This young man was an electronic engineer and was taking those problems, which were beyond the functional expertise of his boss who was a civil engineer. He was advised to work through the strengths of his boss. After a few months I met this young man on the shop floor. He had mentioned that his boss was now treating him as a younger brother. Most of the interpersonal problems were ego related, having emerged out of lack of communication and functional loyalties. A lot of time had to be spent with the people informally as well, to identify and resolve the conflicts.

> Most of the interpersonal problems were ego related, having emerged out of lack of communication ...

Dr Gupta an expert in human psychology was appointed to carry out an analysis of each individual including mine. He was very accurate about individual potentials of people and their IQ. He followed me to Cement Corporation of India and ONGC.

Personal problems of people were not receiving the desired attention. One lowly paid employee was under suspension for over eight years. Another employee who had been suspended along with him for the same crime had been reinstated and promoted. This individual was from a poor family and could not muster enough funds to challenge his suspension. The case was examined and the individual reinstated and promoted from the same date as the other employee, who was his alleged partner in crime. A culture for concern for employees to meet their needs, dreams and welfare, was created.

The 3 Fs

There was a strong need to identify the loopholes in the system and bring out effective changes in the work environment. A culture of receiving anonymous letters had been prevailing for some time. Through deliberate efforts, the person behind this mischief was identified and caught red-handed and the problem eliminated.

A meeting with the office bearers of about seven unions of HEEP was held in the backdrop of the work culture which was established in CFFP where everyone, including the union leaders, had to add value. The two factories together had over 10,000 people. One of the union leaders mentioned in the meeting that now they were aware of the use of rod. I explained that I did carry a stick, an old habit from the army, but had never used it, except in emergency. They were aware that one particular day about 30 security staff (ex-servicemen) of CFFP were dismissed due to misbehaviour of one of their leaders. However, all of them (except two of their militant leaders) were reinstated after a few days on receipt of an apology letter. Hence, a culture of friendliness, fairness and firmness (3Fs) which existed in CFFP was also established in HEEP.

Time Management

It was also noted that there was total lack of concern for punctuality at meetings. The time for the first production meeting was 5 p.m., hardly 40 per cent of the participants were present. To instill within the employees the value of time, latecomers were not allowed to enter during this meeting. This gave the desired signal to the people to maintain time for every activity.

In one of the production meetings, acrimonious discussions took place between senior functionaries of the production and engineering departments on an insulation shop issue. They were asked to meet me next day on the shop floor. Each group brought almost 11 executives, as if they had come for a sport function. During the discussions, one from each group, was identified as the trouble maker. They were asked to meet me in the office. Normally no one is ever made to wait. In this case, they were deliberately made to wait for over an hour in the visitors' room. After that one hour they came into my office and said that they had solved the problem on the issue. Their hour-long confinement ensured forced communication between the two to find the solution. They were also made aware of the consequences of failure to reach an agreement. At this juncture, a serious conflict of group loyalties between the production and engineering departments was also noted. The conflicts were identified and resolved with the clear message to all concerned to keep the overall objective of the enterprise in view and the functional group loyalties subordinated. General manager (operations) and general manager (engineering) were persuaded to visit the shop floor together to avoid conflicting instructions.

A public announcement system was initiated to convey to the employees the loss, the organisation suffered financially for an hour of loss production. It was also announced that those who collect salary without adding any value are worse than beggars. To motivate the employees to put in their best, their contribution towards productivity and production were highlighted.

> ! A public announcement system was initiated to convey to the employees the loss the organisation suffered financially... !

Outsourcing Initiative

To reduce costs, all items that could be outsourced were identified. A large number of small sheet metal parts were being manufactured manually at high cost. These parts could be obtained from the market at a fraction of the cost.

A number of ancillary units had already been established but were not adequately loaded due to bureaucratic hurdles within HEEP. In one of the tea cup meetings within the material department, it was pointed out that the head of the department was mostly busy in meetings, leaving no time for decision-making in the department. Management Services Department headed by a brilliant executive (Mr Gurnam Saran) was asked to look into the problem and restructure the processes for quick decision-making.

The factory had large machine tools for machining of large rotors, casings and other parts. To put psychological pressure on the machinists, material for a number of work orders was placed next to the machines, against the earlier strategy of placing material for one work order at a time. This had the desired pressure on the machinists to work with speed and commitment.

Effective HR Management

One evening I was informed that a conflict with the workers was simmering and the manager of the shop had left for his residence. He was advised to go back to the shop and resolve the issue and report back to me. I waited in my office till he had resolved the issue. This was to give him the moral support but he had to resolve the issue to gain the confidence of his workers. Labour problems need to be resolved on the spot, otherwise they escalate into major problems resulting in loss of time, money and faith.

It did not take very long to create the desired work culture, mutual trust, faith and espirit de corps. My wife took keen interest in welfare activities for the workers and their families including running of schools, hospital and other facilities. A DPS School under control of DPS, Delhi was also established within the campus to provide better educational standards. Sports and games were given major thrust to raise the morale of the community.

Motivational Issue

During one of my daily rounds of the factory, I was introduced to a charge man who had saved a lot of money by reducing rejects by designing a fixture. As per the prevalent culture, his photograph was taken. The charge man however mentioned that such photographs had been taken a number of times. The message was clear that he had never received either a reward or promotion. It is essential that benefit gained by the organisation as a result of individual innovative action must be shared to sustain the interest of employees to use their innovative and creative minds.

Crisis Management—Confrontation between Workers and Security

Central Industrial Security Force (CISF) was introduced in BHEL in line with government directions to guard the factory and other installations in the campus. The workers did not take happily to this change, as they restricted unscheduled movement of workers in and out of the factory. One day as soon as I had entered the office at around 7:50 a.m. I heard a lot of commotion and gun shots from the factory's main gate. The general shift was about to start.

I could not get any information from the staff as they had not yet settled down. So I walked towards the gate. In such critical situations one cannot afford to *wait* for information; one has to take the risk to *move* to the spot of the crisis. One of the general managers, R. Srinivasan followed me. I was surrounded by the workers who escorted me back to the office. They narrated the incident that had sparked off the fight between the workers and the CISF at the gate. They requested me to stay in the office and promised to keep me informed of the situation, which at that time was beyond control. The genesis of the problem was an altercation between a sweeper and a security guard. The sweeper raised an alarm and workers came out of the factory with any implement/iron rod they could get hold of and started attacking the security personnel. The security asked for reinforcements. An inspector with a pistol with some sentries arrived in a truck in the midst of the mob. To frighten the mob he fired shots in the air. As soon as he fired the shots the infuriated mob got hold of him and lynched him to death.

> In critical situations one cannot afford to *wait* for information; one has to take the risk to *move* to the crisis spot.

There were 128 people injured and 22 arrested. The CISF had to be sent to the barracks and the armoury sealed to avoid further bloodshed. The factory was locked out. The gates were managed by workers and officers. Mr Bhatia, the photographer who was taking photographs of the incident from the third floor of the design building was literally stripped and was about to be thrown down, but he was saved by some of his well wishers among the crowd.

The workers demanded that the CISF should be withdrawn and those arrested should be released. They were told that these demands cannot be discussed, let alone agreed to. I however assured them that those who were arrested will be released except those who had started the problem. Discussions were held throughout the night which focused loss of production and the reputation of the plant. At about 4:30 a.m. in the morning it was agreed to restart the factory. Jeeps with loud speakers had already been kept ready to go to the colony and villages around to convey the decision.

At 7 a.m. when I went around the factory at the start of the first shift, the factory had gone into full operation within 10 minutes of start of the shift against 30 minutes, which was the normal time taken on other days. This was achieved by building tremendous trust and faith among the people.

Mr George Fernandes, who was the minister of industry (a former labour leader) refused to accept the facts that the factory had started within 24 hours of such a serious incident and lockout. An IG Police and a senior bureaucrat came to verify. They were surprised to find that the factory was working at full capacity. All possible assistance was provided to the family of the sub-inspector CISF who had lost his life and those who were injured. Extracts from newspaper reports on the incident are given here.

Photographer Beaten Up: Cycles Burnt: Catastrophe Averted by Tactful Handling.

BHEL which witnessed unprecedented scenes of violence and hooliganism on Thursday, resumed normal production. It was due to tactful handling of a very explosive situation by the Executive Director, Col S.P. Wahi that a major catastrophe was averted in this Plant of considerable national importance. ...

Commissioner, Meerut, DIG of Police, District Magistrate, and Police Chief of Saharanpur who rushed to the scene backed the efforts of the management to bring the situation under control. Col. Wahi ordered the factory to be closed down for the day and temporary withdrawal of the CISF personnel. According to the district administration it was Col. Wahi's tactful handling and his superb managerial skill that a major catastrophe was averted. A serious threat to the security of the Plant had existed and if this timely action had not been taken situation could have deteriorated beyond control.

District administration has started investigation pending formal magisterial inquiry. An order under Sec 144 Cr.PC has been enforced banning assembly of more than five persons in the factory area.

Col. Wahi along with the General Manager Mr. Srinivasan and other officers went round the factory to ensure production. He told HT today that everything had returned to normal and there was complete peace. District administration was giving him full cooperation.

Source : *The Himachal Times,* 24 March 1978

*****

BHEL's Black Thursday

The Thursday incidents in Heavy Electrical Equipment Plant of the BHEL, near Haridwar are not only disturbing but have to be viewed from long-term perspective. The Haridwar Plant has had the distinction of being the only unit of BHEL which had worked free from strikes and agitation. The Thursday violence and hooliganism not only disturbed a tradition but also tarnished the image of the Plant and its otherwise peaceful environment....

Col. Wahi's managerial skill saved the situation from adding more painful dimensions to it....

The management has averted the catastrophe but is required to work under heavy strain to bring real peace in justice for fairness to both the parties.

Union Ministry of Industry and the Corporate Office of the BHEL should give full support and freedom of action to Col. Wahi and his officers to heal the wounds and maintain productivity. ...

Source : *The Himachal Times,* Dehradun 28 March 1978

Leadership under Dr Krishnamurty

BHEL had a good reputation of looking after the customers and maintaining good relationship. Dr Krishnamurty, an outstanding chief executive, had set excellent norms for culture and was the role model as a leader. He had given total freedom to the unit heads to operate.

In one of my visits to his office in Delhi, he spoke to me with all the seriousness he could muster, "Colonel we have established very strong systems and procedures and you seem to be flouting them." This happened when I was involved in the construction of Central Foundry Forge Plant. I was aware of rumour carriers. I replied, "Sir, either you can have performance or total subordination to system and procedure; morale and motivation of employees is my responsibility and so is money, the bottom-line of any enterprise." He smiled and said, "Go ahead, I was only mentioning the environment around you." We received full support from Dr Krishnamurty and his trusted lieutenants; an important factor for the success of CFFP.

Taking Risky Decisions

HEEP had an order for 235 MW generators from Atomic Energy Commission for one of their units headed by Mr M.R. Srinivasan (who later rose to the position of chairman of Atomic Energy Commission and thereafter member, Planning Commission). They were insisting on a 'type' test for the generator, a very stringent test which involved very high overloading of the system during testing. Normally such tests are undertaken when the first generator is manufactured after the design is completed to prove the latter. We were manufacturing these generators with Soviet collaboration, who had undertaken the type test to prove the design. Mr Srinivasan would not relent, and insisted on type test. This problem had existed for almost a year before my taking over HEEP.

I had a chat with the senior executives from production and engineering departments. It was decided to conduct the type test. It was a major risk we had taken without even the involvement of the Soviet collaborator and director (engineering) BHEL. We had to take the risk and accept the accountability. The Type test was successfully conducted in the presence of Atomic Energy officers. We had anxious moments till the test was completed.

Lighter Side of Industrial Life

To celebrate the occasion, as mentioned above, we had organised a dinner in the lawns. I was asked to say a few words on the mike. I said, "I am drunk—(a pause) with the pride of my people who have achieved success and set the pace for closer collaboration with the customer." Next day the rumour managers circulated that everyone was drunk! The executive director had admitted this himself in public! This is the lighter side of hard industrial life in India.

Visit by Icon

The general secretary INTUC (Congress Party affiliated) Union approached BHEL for accommodation of Mrs Indira Gandhi (former prime minister of India) at its guest house during her visit to villages around Haridwar. She intended to see for herself the atrocities committed on backward classes.

The visit was promoted by Mr Garg, a Supreme Court lawyer who was president of one of our unions in BHEL. He had shown his bloodstained clothes to Mrs Gandhi, as a result of the blows he had received when he had gone to protest against the atrocities committed against backward classes. (It is rumoured that these blood stains were created by killing a goat.) Mrs Gandhi came to the guest house in an ambassador car with six occupants. She was sitting in the front seat along with Mrs Moshina Kidwai and the driver. She stayed overnight in the guest house and left next morning without any fanfare. One developed tremendous respect for her ability to sacrifice even personal comfort to fight for the sake of the depressed people. I had the honour of escorting her twice earlier, when she had visited Chennai to inaugurate the Heavy Vehicles Factory, Avadi, and later, for the inauguration of the first blast furnace of Bokaro Steel Ltd.

Empowerment

We had hired the services of a brilliant behavioural scientist late Prof. Nitish De. He worked with the people in the fabrication shop. He was able to generate the desired enthusiasm among the workers for job enrichment and work redesign. The workers enjoyed the empowerment received. They themselves reduced the manpower including supervisory staff. Each artisan acquired multi-skills to reduce idle time and to improve productivity. The productivity improvement was in some cases more than 35 per cent. A culture of higher productivity was generated in both the plants which improved the performance manifold, both in terms of productivity and production.

Transparency in Negotiations—Contract with KWU (Siemens)

BHEL management decided to arrange a collaboration with KWU (Siemens) for the manufacture of 500 MW thermal sets. Their top management had visited Haridwar to see our facilities. A delegation of BHEL executives including myself and headed by Director (Engineering) Dr Saran, was deputed to visit their works in Germany. After visits to their works, a presentation was made to plan the schedule for technology transfer as well as prices. I got a feeling that they had no intention of transferring about 20 per cent of the technology. The prices were more than the contracted price BHEL Haridwar was to get from the customers. I made my reservations clear during the discussions, though I had supported the induction of Siemen's technology.

The contract was signed at the BHEL headquarters at Delhi, who were in picture of the total contract clauses and its impact on the commercial aspect of BHEL operations.

Dealing with Parliamentary Committees

This contract with Siemens (KWU) became the subject of examination by COPU (Parliamentary Committee on Public Undertakings). I was called in as a witness. The committee chairman warned me about the importance of the committee and my obligations to be respectful to it. I was asked to reply to the questions to the best of my remembrance. The first question was asked by Mr Subramaniam Swamy, "How did the prince (he meant late Mr Sanjay Gandhi) come to Haridwar?" I said, "I shall tell you how he came." Mr Subramaniam Swamy interjected, "He came by helicopter." I said, "He came by car." There was a momentary laughter. I proceeded to explain that he had come with the Chief Minister Mr N.D. Tiwari along with many others.

How was the contract with Siemens finalised was the next question. I explained that it was finalised at the BHEL headquarters. One member said, "You call your self executive director and you know nothing." I immediately addressed the chairman and said, "The honourable member is obviously not aware of BHEL organisation. The BHEL headquarter is at Delhi whereas I am located at Haridwar." The chairman apologised on behalf of the member, and after a few more questions and answers not related to the issue, I was asked to retire. As soon as I came out, a Sikh officer who was assisting the committee came to see me off and said, "Not many people who come to this committee go out without crying. Congratulations you had made them cry!" I was aware of my commitment to BHEL, my old organisation, which had been led by an outstanding chief executive (Dr V. Krishnamurty).

Training and Development

A culture of development through training programmes was established to improve the morale of the employees and motivate them for growth. A number of my junior colleagues had achieved distinction and high positions later in life. To mention a few are—Mr P.S. Gupta general manager (operations) and Mr Gavisadapa senior manager both of whom had reached the position of CMD of

BHEL later; Mr R. Srinivasan general manager (personnel) joined ONGC as member (personnel), Mr K.N. Khanna, general manager (finance) became director (finance) BHEL and had also officiated as CMD BHEL. Mr Saran Dy G M (MS) had reached the level of director BHEL. Mr R.S.S.L.N. Bhaskarudu, a manager reached the position of managing director, Maruti and later chairman, Public Enterprise Selection Board. Two of these executives had been transferred to Haridwar

> ❗ ... people can perform even beyond their normal potential, if provided with proper work environment... ❗

from other units of BHEL, under not very happy circumstances. Both of them had outstanding technical knowledge and helped BHEL Haridwar to create additional high-tech facilities. Their potentials were fully exploited for the benefit of the organisation. Such experiences bring out clearly that people can perform even beyond their normal potential, if provided with proper work environment to ensure high morale and motivation.

Change of Government and Its Associated Problems for Managers

The Janata Government had come to power in 1977, on the defeat of the Congress. I received an anonymous letter containing frivolous and fabricated acts of omission and commission from the ministry, with remarks of George Fernandes, Minister of Industry. The main complaint against me was that a number of executives who were related to high dignitaries had been recruited. A lot of money had been spent during the visit of Mr N.D. Tiwari (chief minister of Uttar Pradesh) who was accompanied by Mr Sanjay Gandhi. The minister of industry wanted reply 'blow by blow' which was sent. Within a week a few more queries were received. This looked like an endless process resulting in wastage of time and effort. I requested CMD BHEL to arrange my meeting with the minister which was granted immediately.

At the appointed time I was received very coldly. Even my greeting was not responded and even a chair was not offered. I pulled the chair and sat down and addressed the minister, "Sir what is your problem?" He looked up from the file which he was ostensibly trying to examine and said, "Colonel I thought it was your problem." "Sir, I have no problem. I am managing two factories with over 10,000 personnel. There are bound to be a few bad hats." He immediately responded, "There must be a few Bs." I replied, "I did not want to use that word and since you have used it, I agree. This is the work of those Bs. I am a professional manager. I have no allegiance to any political party but have respect for people irrespective of the political party they belong to. In case you want a two-line letter, I have brought it with me." After this he spoke to me for about 30 minutes and explained his views on management and assured me that no one would bother me any more.

I was positioned as chairman and managing director of Cement Corporation of India in schedule-A (the highest in public sector) as personal to me, as the enterprise was still being considered for upgradation to schedule B from schedule C. The 'rumour managers' were not aware of this fact, otherwise I could have been labelled as a close associate of George Fernandes. He did give full support for the revival of cement corporation. A clear-headed outstanding administrator politician.

2

Strategic Planning – Cement Corporation of India

I joined Cement Corporation of India (CCI) on 26 October 1978 as chairman and managing director in schedule-A. Cement Corporation of India was having indifferent performance, even though it was being managed by the so-called cement experts. The plant at Bokajan (Assam) was closed because of a strike for over a month. No director had visited the plant to resolve the issue. The morale of the employees was low. The projects at Yarguntla (AP), Neemuch (MP) and Akaltara (Chhatisgarh) were behind schedule. The operating efficiency of plants at Mandhar (MP), Karkunta (Karnataka), Bokajan (Assam) and Rajban (AP) was much below capacity. A number of good experienced people had left the organisation and some were on the verge of doing so.

The ambience of the headquarters at Delhi indicated low morale, lack of motivation and indiscipline. There was no sense of time management. The previous CMD, who I understand was an excellent cement expert, had taken voluntary retirement to set up his own cement factory in Andhra Pradesh. The corporation was being managed by director (operations) Mr A.P. Maheshwary under the supervision of a group of secretaries to the government of India. Engineers India had been appointed as consultants for projects for expansion. Cement Corporation had picked up a bad reputation for low productivity, delayed project completion and financial losses. It was on the verge of being declared a sick company.

A meeting was held with the executives at the headquarters to find out their views for the dismal situation and evolve recommendations as remedial measures. It was like a situation audit resulting in SWOT analysis. Excellent views were expressed and noted. These were substantiated and supplemented by visits to the plants and projects.

Valuing Time Management

I explained to the executives my expectations to make the Corporation a model organisation in the cement industry with correct values and ethics. Particularly, time management in every activity was stressed, starting with the executives to set a personal example in coming to office and attending meetings on time. They were also advised to stage themselves as role models for hard work and commitment. I had the habit of talking to my senior colleagues, first thing in the morning and at the end of the day before leaving the office and expressed my intentions to continue with it. I however found that the director (operations) did not take the advice seriously. On two days, he had come late. On the third day, I got his office locked and he was obliged to come to me to get the key.

Evaluating Problems

On the very first day of my having moved into my new office, a fire accident had occurred which had destroyed most of the CCI headquarters. The accident happened at night, a few hours after I had left the office around 6.45 p.m. Fortunately for us, only a day before we had the office insured by our Director (Finance) Mr Kapoor. I was informed of the fire by the Company Secretary Mr Venkatraman and had rushed to the office to supervise the fire fighting efforts. An appeal (see below) was generated and circulated by the union leaders which brought out the level of maturity, good intention and faith that the workers possessed for the Corporation.

Appeal of Workers

An unfortunate fire incident in our Corporate office has destroyed most of Headquarter's valuable records and equipment. The loss suffered is known to you. All this has happened at a very critical stage. Therefore, the challenge thrown on us by this event has to be faced by all of us with courage and determination.

The reasons for the said accident is under investigation by the authorities concerned. Kindly give only correct information if asked by the union representatives, officers and investigating authorities and help us to stop the spread of rumour. The effort made by all of us under the dynamic leadership of our directors and the Chairman and Managing director Col S.P. Wahi for building a respectable position of the Corporation have been appreciated by all sections of the society....

The setback caused by the event is not small. We have to cover the gap with our sincere, dedicated and hard work in a period as short as possible. We are confident that we are capable of doing this. We have therefore to forget our minor problems and have to first take care of the only problem of re-establishing ourselves and to reach at the advanced stage that we were. For achieving this, united efforts of all of us are needed.

Honourable Minister for Industry Shri George Fernandes who visited the office on 21.5.1979 has also assured full government support needed by us. Therefore, when the union and the management are united, and full government support is available, the only thing wanted is good leadership which too is available. Therefore, it is appealed to all to pledge as under:-

1. Reach office and start work before office hours.
2. Do more work than allotted to you.
3. Respect your elders and seniors and love your juniors and inspire them. Motivate your colleagues to take initiative, accept additional responsibility and adopt positive attitude.
4. Avail less than the normal lunch time, work even after office hours and holidays and do not claim for the same.
5. Avoid absenteeism and awake the fellow employees to follow the common goal and objectives.

Kindly note that this is the time for us to work hard. Do not bother for minor problems and your rights.

Be vigilant.

Sd/- G.S. Nair,	Sd/- J.P. Sharma	Sd/- S.K. Taqi
Secretary	Vice President	General Secretary.

The real problem lay with the leadership at various levels. Self-interest was a dominant factor in their decision-making and actions. In most cases the top management was at fault. Here, the two seniormost directors were the pace-setters in the wrong direction.

The situation was corrected through organisational changes and delegation of work down the line. Management through 'wandering' ensured involvement of rank and file and fast feedback for corrective measures. An extract from *India Tidings* (10 June 1979) has been reproduced here.

Col Wahi—the Man behind CCI Success

A few months ago, a SOS went to Col Wahi to come to Delhi and take up the Chairmanship of the Cement Corporation of India which was in shambles. The first thing he did was to draw up a ten-year perspective plan involving an outlay of Rs 10,000 million and taking production to dizzy heights. He had also decided to centralise the CCI office in Nehru Place and shifted the entire facility on three holidays. Perhaps this was too much for the Gods. Or could not the enemies' stomach the achievement of this man machine? Whatever it is—this has to be decided by a probe body—on the night of May 19 the offices of CCI went up in smoke. Col Wahi was there rushing up and down trying to salvage whatever he could. Though initially depressed ('My own house has gone up in smoke') he lost no time and in a gap of 24 hours set up CCI.

The Minister of Industry, George Fernandes, who is giving Wahi his utmost support visited the place and went up and down eight floors—the lift had also become a casualty—and pronounced that the CCI shall rise. It has risen like phoenix.

Col. Wahi cannot remember any enemy because he is incapable of making any. Even on the night of the fire his thoughts turned only towards friends. The friends came in drives from IFFCO, SAIL, NHPC and others. IFFCO's Managing Director, Paul Pothan offered his office in a different block. Dr P.L. Agarwal, Chairman of SAIL, came with a consignment of typewriters. The executives of the National Insurance Company and the Oriental Fire and General Insurance were at hand to give all help. Result: none who knows about the tragedy will know that the CCI had another office which was burnt down.

Col Wahi has made one thing clear by his phenomenal assembling of the CCI. He will not be cowed down by threats or arson. CCI will march on. His detractors would have to meet him in fair combat, not on the sly.

Tackling the Bokajan Crisis

I visited the Bokajan factory which was on strike. The union leader and his wife, who also was a union leader, met me along with other members. I heard their tales of innumerable problems affecting the efficiency of operations, and the welfare of the employees and their children—lack of proper means of transport to the mines, inadequate housing, recreational and educational facilities, indifference of the organisation towards these problems and the resultant low morale of the workers.

At Bokajan, the union was assured of getting all the issues settled to create the desired infrastructure, and provide necessary welfare, sports and educational facilities. Orders were issued on the spot for construction of houses and other facilities. The unions wanted me to address the striking workers, to which I replied that I shall address only after the factory is restarted. I also explained to them that the factory will be started the same evening even if workers are not available. The kiln was started the same evening by the supervisors and the officers. Thereafter the factory did not have another labour problem. It achieved more than 100 per cent capacity utilisation against 20 per cent in earlier times.

Instilling Values for a Positive Organisational Culture

Initiative to achieve value addition was non-existent. The plant was producing a very high grade cement and selling in the market at the price of pozzolona cement. The traders were adding pozzolona material to Bokajan Cement and making huge profits. The situation was corrected when the issue was brought to focus. During one of my visits to the marketing department, I had found a large number of empty liquor bottles. This gave an indication of the prevalent culture, which was corrected through firm handling and institution of proper controls. The induction of Mr A.L. Kapoor as director (finance) was a great help to streamline the working. Mr Kapoor had worked earlier with me in BHEL Haridwar as deputy general manager (finance). He set the pace for integrity and values which we shared for proper governance. A few other young executives like Mr Arun Dutta and Mr S.M. Dewan were inducted to be the role models for hard honest work and commitment.

Constitution of Study Group

The ministry of industry constituted a study group (vide order dated 26 December 1978) to review the working of CCI and its organisational structure and ensure that *a)* the ongoing schemes are completed with expedition; and

b) production units are managed efficiently.

Under my chairmanship, Mr T.V. Mansukhani, Prof. Nitish R. De and Mr Sher V. Utamsingh—were appointed members of this group.

Prior to the appointment of the study group, certain actions, as listed in the following paragraphs, had already been undertaken, resulting in a dramatic improvement in the morale and motivation of the employee and the performance of plants and projects.

Employee Motivation

As indicated before, the morale of the employees was very low. They were being treated as mere numbers. There was no forum for formal identification of employee problems except through unions, which resulted in confrontation. Informal interaction with the people was missing. They were not included in any decision-making process. All decisions were arbitrary which had an adverse impact on the morale and motivation of the employees, and in consequence affected the financial bottom-line of the enterprise.

The higher rungs of management were only busy in purchases and sales. There was a total lack of support, either administrative or professional to the plants/projects. The executives were stagnating in their existing positions for years. A number of good ambitious executives had left the organisation and a few were on the verge of leaving as they felt that the future of CCI was bleak. Those who were left, either had a secondary source of income or had reconciled to their fate. The dismal situation needed immediate actions to remedy the state of affairs.

During visits to the plants and projects, I interacted with a cross-section of the employees including the unions and got some excellent ideas, to create a future for the organisation and employees.

EIL Consultancy

A month or so prior to my joining CCI, Engineers India Limited had been appointed as consultants for new projects, under the instructions of the government, even though EIL had never provided consultancy on cement. Owing to their overpowering intervention and inadequate knowledge, the EIL Management failed to motivate the executives. The EIL management was asked to make a presentation before me on the expertise available with them. In the process they accepted their inability to handle our assignment and had to refund most of the money already paid to them. Then CCI opened its own engineering consultancy cell with the existing people and a few inducted from the market. This group was able to engineer the new projects at Adilabad and Tandur and provide excellent support to the projects under construction. This also helped to boost up the sagging morale of the CCI executives, who were being harassed by the EIL Manager in charge of the cement section without even having the elementary knowledge of cement plants.

> ! They could see their own growth with the growth of the organisation. !

Long-term Vision

A ten-year perspective plan was drawn up with in-house expertise, which was plenty but had not been utilised in the past. The plan was widely circulated within and outside the organisation, including the planning commission. It created the desired enthusiasm among the employees. They could see their own growth with the growth of the organisation. The private sector saw this plan as a threat. Some of their good employees started applying for jobs in CCI. Apart from the projected growth we had changed the designations of employees, particularly the plant managers who were re-designated as general managers. We were the pace-setters in this regard. This small change in designation prompted a number of bright managers from the private sector to apply for jobs in the CCI. Word had already spread about the rejuvenation of CCI with phenomenal possibilities of growth.

The projects/plants were given adequate financial and administrative powers, so that they did not have to look over their shoulders for help. A culture of trust and mutual respect was established. Through performance budgeting, proper feedback and controls were established. Communication channels through telephones/telexes were established to give immediate support from the corporate headquarters when needed.

Human Resource Development

A training institute was established at the Neemuch Plant to ensure continuous training of employees on every aspect of work including safety and environment. A number of executives were sent abroad for training. All these actions had an electrifying impact on the performance of the plants and projects. The projects started getting completed at an accelerated pace. The plants achieved excellent capacity utilisation in spite of constraints of power shortage and poor quality of coal. In some months when power position was better the plants achieved more than 100 per cent capacity utilisation. This proved beyond doubt the potentials of people to achieve results which otherwise are considered humanly impossible unless proper working environment is created and the morale and motivation of employees ensured. Leadership abilities are more important than managerial skills, though both are essential to be developed in managers. To ensure workers' commitment, leadership has to set personal examples. Constant communication with the employees has to be maintained about challenges. A message to the employees is appended below as an example.

Message from the C & MD

I am making use of this opportunity to share with you some of my thoughts about the future of our Organization.

On the growth of an Organization depends the growth of its people. Our Perspective Plan is under very active consideration of the Government and a number of Projects are likely to be sanctioned. We are in touch with the World Bank and other financial institutions for necessary financial assistance.

The Organization's structure has been revamped for meeting the present challenges and future growth.

The quality of manpower resources is our best asset. Its optimum utilization by participative culture, backed by well-planned systems and procedures has to be ensured. A lot of work has been done in this direction, which is already showing the desired results.

The development of personnel is receiving due attention. A number of training programmes have been conducted and a training institute is being set up at Neemuch. During my recent visit abroad, the serious gap in knowledge and operational efficiency in our units as compared with the units in USA, UK and particularly, in France was noticed. We are taking deliberate measures to ensure that a large number of our people are exposed to the latest developments in the developed countries. We can ensure substantial economies in the construction of our plants and their operations by following this strategy only.

Besides, action is called for in a number of other areas, particularly, in-house R&D facilities, centralized workshop, maintenance and spare holding to enhance internal capabilities for optimizing our operations and ensuring economy in operations as well as to keep pace with our future growth.

Keeping in view the economic growth in the country and the scale of construction activity in the next 10 to 20 years, preparatory measures will have to be thought of for bulk handling of cement in association with Ready-Mix Industry. Due emphasis has to be given to diversification programmes to improve the financial position of the company.

The tasks ahead of us are really challenging and we have to match this by continuous improvement and development of our human resources.

A drive is being made to improve the educational and medical facilities at the plants/projects. I am happy to note that every member of CCI Family is showing desired sense of belonging and is striving hard to improve the operations.

May I also take this opportunity to wish all of you and the families a Happy Diwali.

Profit Sharing

Cement Corporation of India in 1979-80 showed a profit of Rs 109.85 lakh against a loss of Rs 82.46 and Rs 98.31 lakh in the previous two years. In view of the cumulative losses, the workers were entitled to only statutory bonus of 6.23 per cent. They were entitled to this bonus even when there were losses. It was decided to pay them bonus in the form of productivity linked incentive scheme by lowering the standards. One cannot deny the workers the fruits of their hard and dedicated work. This action further increased their commitment and there was a significant improvement in the next two years.

During one of my visits to Rajban (HP) factory, a file was put up to me to sanction the movement of limestone from the mines which were 10-15 km away from the factory by road, as the ropeway supplied by Jessop and Company was not operating well. Discarding the suggestion I wrote on the file: 'Over my dead body'. **The ropeway started working and all other so-called technical problems in the operation of the plant got removed in a matter of days.** The plant was supplied by Larsen and Toubro, one of our best manufacturers. **Most of the technical problems are man-made due to human relation issues—either ego or greed.**

Networking with the Government

With the improvement in performance through visible commitment and hard work of employees, the Corporation received excellent support from the government and the planning commission. This enabled the Corporation to show excellent results. The CCI study group had taken note of the actions which had been initiated since October 1978, and incorporated that in the report which was submitted to the government on 31 March 1979. The report also covered upgradation of the Corporation as well as creation of a few senior positions. These recommendations helped us a great deal in getting full support from the government. I also visited cement plants in UK (Blue Circle), France (Lafarge) and USA and submitted a number of recommendations to the government. On request from the two state governments, I was asked by the government to be on the boards of Malabar Cement and UP Cement Corporation.

Bharat Opthalmic Glass Ltd

The government showed their trust by asking me to accept an additional charge—Chairmanship of Bharat Opthalmic Glass Limited at Durgapur, for a year, to set the pace for this sick organisation on the path of improvement.

This plant was set up with the collaboration of the Soviet Union. I visited the plant a number of times. Discussions were held with the unions, executives and individuals who had been trained abroad. It was noted that apart from the obsolete technology, the real problem was the issue of human relations. Durgapur was infamous for labour problems. The first day when I visited the plant, I could see workers watching my movements from the windows, along with a few executives. I discovered later that managers were scared to visit the plant for fear of physical confrontation with the workers. There was total lack of discipline. The employees who had been trained in USSR had either left the organisation or were not employed on the shop floor. The meeting with the workers revealed that they were ready to cooperate but the managers had got used to doing nothing due to fear of worker's indiscipline. With organisational change and pep talks with the workers and executives, the performance showed signs of improvement. A healthy collaborative relationship was developed between the workers and the management. I then advised the government to appoint a full time CMD.

> A healthy collaborative relationship was developed between the workers and the management.

Cement Research Institute

The government also asked me to chair a committee to look into the working of Cement Research Institute. The institute had excellent systems and procedures and was staffed with brilliant scientists. It was, however, burdened with innumerable human relation problems, mostly created by the top management because of selfish agenda at the cost of the organisational health. A report was submitted to the government, bringing out the necessity of training managers on human resource management and leadership abilities.

Change in Government

In the meantime, there was a change in government. A nationalist government which had got the freedom for the country came back into power. The ministry continued to be helpful and CCI continued to progress. When I left the cement industry to join the oil sector, CCI had by then seen a phenomenal growth. It had become the role model for the cement industry.

On my appointment at Oil and Natural Gas Commission (ONGC), I was asked to hand over the charge to a man who I knew was incapable, and this in spite of my adverse confidential report on him. It did not take long to bring down the operation of this good organisation which had become a role model in systems/procedures, productivity and human relations. The organisation went downhill due to poor leadership at the top. My successor had to be unceremoniously removed soon.

My Tryst with Black Gold

10

Oil – the Life Blood of Economy

It was my long cherished dream to be associated with the oil sector which was fulfilled by my selection to the position of chairman, Oil and Natural Gas Commission (ONGC) by the Public Enterprise Selection Board (PESB). I was inducted in ONGC in June 1981 as OSD (officer on special duty), to ultimately replace the incumbent chairman, a distinguished finance expert, who was to retire on 1 October 1981.

It was my good fortune to have been at the helm of affairs of ONGC during most part of the 80s, which in many respects was an eventful decade for oil—the fall of the Shah of Iran and taking over by Ayotallah Khomeini; the Iran–Iraq War of 1980; and the 5th post war oil shock saw the oil price shoot up to US $34 per barrel. This however did not prolong much since more and more non-OPEC oil started flowing and so was the case in India where oil from the newly discovered 'Bombay-High' was available. The race for supply and a price-war between OPEC and non-OPEC countries resulted in a price crash to as low as US $6 per barrel in 1986. This low price resulted in reduced exploration activities worldwide but we utilised this opportunity to expand our exploration activities both onshore and offshore and established additional reserves at a low cost.

The pages to follow **detail the strategy and tactics followed to mobilise the inherent strengths of the organisation to identify and grab opportunities to achieve self reliance in oil at an affordable cost.** In fact, the seventh plan of the country was publicly acknowledged as the ONGC Plan, which saved the country's valuable foreign exchange.

During the period of a few months, as OSD I was able to visit each work centre spread all over the country, and the three R&D

institutes that existed at that time in ONGC. Visits to some oil fields and R&D institutes in former USSR, to the UK part of North Sea, and interaction with CFP, a subsidiary of TOTAL (an integrated French Oil Company), who were consultants for the development of Bombay High oil field brought to focus the exciting challenges ahead in the field of energy, particularly oil exploration and production.

I realised that oil exploration is a scientific gamble. It does not stand by input and output engineering ratios. Here input is deterministic but output is probabilistic. It is a game for bold leadership coupled with high perseverance capabilities.

Oil – Life Line for Economic Growth

After the Second World War (1939-45), oil became the single most important source of energy, and also a vital industrial input particularly for the transport sector. The formation of OPEC (Organisation of Petroleum Exporting Countries – Appendix J) in 1960 and subsequent oil price shocks, brought to focus the use of oil as an economic and political weapon. The unfortunate conflicts in Middle East and events in Iraq prove the point. Pandit Jawaharlal Nehru—the architect of modern India was a thinker and a visionary. Much before Independence, while he was lodged in jail with K.D. Malaviya, he was day-dreaming about the country being self-sufficient in oil. He had once said,

> I have spent 19 to 21 years in jail and K.D. Malaviya was also in jail with me. We had then a definite purpose which kept our minds engaged—we must find oil in our country and improve our financial position and be self-sufficient.

After Independence in connection with the formation of ONGC, he declared in Parliament on 26 May 1956:

> Oil is of vast importance in the World today. A country that does not produce its own oil is in a weak position. From the point of view of defence, the absence of oil is a fatal mistake.

Mr K.D. Malaviya as minister in the Indian cabinet gave practical shape to the dream of Nehru. He had to overcome many obstacles both internal and external but through his commitment and perseverance he succeeded in his mission and is rightly called 'Father of Oil in India'. He was an extraordinary man with tremendous foresight and vision for the future. His following statement indicates the motivation for his determination to put India on the world oil map.

> Success in finding considerable quantities of oil will make a tremendous difference to India and India's economy. We have seen in recent months and years how vital is oil, not only in the world's economy but in world politics.

> ! I, as a professional manager, perceive Indira Gandhi as a leader of great vision and wisdom. !

Fortunately for India, other leaders who succeeded Nehru's era had also a somewhat similar determination and vision. Indira Gandhi, as prime minister of India while naming the institute of petroleum exploration in Dehradun as K. D. Malaviya Institute of Petroleum Exploration, in December 1981, had remarked,

> We cannot regard oil merely as an essential commodity but have always to keep in view its political sensitivity and the consequences of any action connected with it.

> The foundation of our programme for oil was laid by my father's vision. His vision and his dream were transformed into reality by determination of Shri K.D. Malaviya, others like him and the band of technologists whom he had inspired. Memories of Shri Malaviya's dynamism and his dedicated persistence in the face of tremendous odds are fresh in my mind. Until the very end, he came up and pursued ideas not only for the proper utilisation of oil resources but a new energy like fusion. This is an occasion to reflect on the larger aspects of the question which motivated our indigenous oil effort when the going was really difficult and the scene was dominated by multinational oil companies.

I, as a professional manager, perceive Indira Gandhi as a leader of great vision and wisdom. She was very well informed about the oil and the strategies followed by other countries to manage this vital depletable source of energy. She had a deep understanding of the probabilistic nature of oil exploration and had known the strategy being followed by countries like USA. It was this realisation that made her ask me besides other searching questions whether we should exploit all the oil fields as and when we discover them. This question is even more relevant today from the point of energy security. She had rightly appreciated the strategic importance of oil to be retained under the control of the public sector. This strategy was further reinforced by Rajiv Gandhi. Unfortunately the events and actions of early 90s (I had left ONGC in December 1989) moved in a different direction and resulted in a decade of practically no addition of oil and gas reserves in the kitty of ONGC. This has been discussed in some detail in later chapters.

Birth of ONGC

It was the fruit of Nehru's vision and Keshav Dev Malaviya's commitment, which ensured the establishment of ONGC by an act of Parliament in the year 1956. Malaviya had to struggle to get technology and equipment from abroad, against the lukewarm interest of the Western world. Their oil companies had been operating in India freely till 1947, but in spite of their financial and technological muscle, they were not able to discover oil in any part of the country except Assam. Their expert opinion was that no more oil could be found in other parts of the country. In fact, one of these experts is alleged to have said, "If you find oil in Gujarat, I shall eat my boot". We have been looking for that expert since the discovery of oil in Gujarat, to enable him to redeem his commitment.

My Tenure at ONGC

During the initial period of my stay in ONGC as OSD, my visits to various work centres and interaction with the people at all levels brought the challenges ahead of the organisation to transform and to harness full potentials of the people. The exploitation of the huge hydrocarbon resource base was documented and prioritised with the help of ONGC executives. A few enablers were identified as action had to start in many fronts to make an impact to energise the people. A number of innovative steps were taken to galvanise the spirit of the people and correct certain imbalances that had crept in the organisation. Environmental concern and human resource management became the greatest anxiety of the organisation that gave an all-round improvement and satisfaction never witnessed before. **In almost all the fields, whether it be surveys, drilling, production or accretion of oil and gas reserve, phenomenal peaks of achievement were obtained.** The following pages detail the issues that were perceived at that time which called for action, and the initiatives that were taken to bring about the change over.

SWOT Analysis—ONGC

1. Lack of Organisational Base

The greatest strength of the organisation was the inherent potential of the people who had been well trained in the technical functions/disciplines abroad, mostly in USSR. At the same time, the organisation that had a 25 years standing in 1981 when I joined,

> People were stagnating in their positions for years with bleak future for growth.

had not received much attention towards organisational development. All members of the Commission including the chairman had been inducted from other organisations. Not a single individual from the first batch of executives (1956) had risen to the board (commission) level. There were only six general managers (two in finance, one in institute of drilling technology, one in institute of petroleum exploration and one each in offshore production and offshore engineering construction). It is obvious that in this scenario there was no possibility of selecting any member or chairman from such an organisational base. It was noted that organisational development had received scant attention in the absence of a long-term corporate plan. People were stagnating in their positions for years with bleak future for growth. Scientists and engineers had been given good initial functional and technical training, but managerial and leadership development on organised basis was missing. The continuous upgradation of technical and functional knowledge, so essential to achieve excellence in performance, was also not in focus.

The development of internal talent to reach the top echelons of management, so vital for the management of a highly technologically oriented organisation, had not received any attention. The only technocrat, Mr Negi, during whose tenure major successes in oil discoveries including Bombay High had been made, was not even confirmed as chairman, though he had officiated in that position for over four years. There was such a low focus on organisation, whose core business was oil exploration that it did not have a whole time exclusive member (exploration) at the board level for almost 10 years. No clear mission and objectives were available to ensure sustained and coordinated working. It were the efforts in keeping with the commitment that I had made at the time of joining ONGC that by December 1989 at the time of my retirement, every member in ONGC had grown within the organisation.

2. Bureaucratic Working—High Severance Rate

At the time of my joining, the organisation was being run on bureaucratic lines. Recruitment and promotion policies as laid down in recruitment and promotion regulations of ONGC were vacancy oriented. To create one vacancy a major exercise in paper work through various echelons of the organisation was involved. The morale and the motivation of the people was obviously very low. The severance rate of highly trained executives from reservoir engineering, production engineering, instrumentation and drilling was very high, mainly due to the recruitment by the Middle East countries. This was a matter of serious concern.

3. Non-Functional Corporate Headquarter

The culture at Dehradun, where more than 3,000 employees and two R&D institutes were located, was like LOB ('Left out of Battle'). There was no sense of involvement, total lack of discipline and freedom from accountability. Dehradun in those days did not have an air link and movements were only by rail and road that was slow, helping the sleepy culture to prevail. Dehradun did not receive the importance it deserved as a headquarter/corporate office.

Having noticed this situation, I had moved my family to Dehradun within a week of assuming charge. This was essential to bring about some system and subtle importance into the management of the corporate headquarters.

4. Communication Systems

The communication system was antique except in Bombay offshore. There was not even a telephone hotline between Dehradun and Delhi. Hot lines were established only in times of crisis, but would soon fell into disuse once the crisis was over. The link with Assam and other remote places was through radio which was mostly silent. The information flow was through postal services and long telegrams received sometimes after 15 days which at best could be used only for postmortem. Live advice and support which a corporate headquarter is supposed to provide was missing. The corporate headquarter with such a state of affairs only functioned as an audit party and did not 'lead from the front'.

5. Disparity in Work Cultures—Onshore–Offshore Divide

The then organisational structure, based on offshore and onshore, had created a serious divide between the two groups resulting in sub-optimal utilisation of resources and had created two distinctly opposite cultures. Offshore activities were either totally or partially done by foreign contractors whereas the onshore activities were handled in-house. As a result there was visible personality difference between offshore and onshore executives. Offshore executives looked white collared against their blue collared onshore counterparts with associated complexes in behaviour.

Association with foreign offshore contractors brought talks in terms of dollars in a country where Rupee is the official currency. The culture of 'do it yourself' gave way to 'get it done'. Proximity with the American companies for offshore and dependence on self for onshore had its own cultural effect. In other words, the induction of new technology offshore brought along with it not only new values and a new economy to the organisation, but also undesired cultures. Under such a scenario, the onshore activities could not afford to remain isolated from these new ideas and new economy without peril to the overall development of the organisation. At the same time, the valuable work of the scientists of the R&D institutes, funded by the organisation, received no or scant attention offshore. This had to be integrated. In effect a resolution (or should I call it a revolution) was required to break down the barriers and re-organise to absorb and adapt for the impending 'knowledge revolution' round the corner. In retrospect, I am glad that this was a right decision.

6. Equipment and Materials Management

Materials management was in a state of disarray. More than 100 crores of imported equipment were lying in Calcutta and were being periodically moved from one warehouse to another. The equipment and spares had not been codified and some equipment were being ordered over and over again. There was nothing like inventory management in the absence of proper documentation. In onshore areas equipment was antique. The computer system at the headquarter was obsolete and there was lack of computerisation in the organisation. Out of 34 land drilling rigs only 20 were in operation. The rest were lying idle and mostly in a state of disrepair.

Two rigs were lying at Jodhpur railway station in cannibalised condition. The member (material) at board level was only involved with purchases and was jokingly called member (immaterial). In fact this triggered in my mind the need to re-organise the system and the working of the board level executives.

7. *Corporate Governance*

Corporate governance at ONGC was controlled by an Act of the parliament. Most of the times, the delegated power was negative and one could not act efficiently by following it to the letter as it presented many administrative hurdles. One had to go to the government to give any financial benefit to the employees. The organisation was under the control of desk officers of the Government of India. During my tenure of over eight years as chairman, I had to deal with five secretaries and seven ministers in the ministry of petroleum and natural gas. I managed to survive by maintaining the highest standard of values, upholding the correct protocol and with firm handling. Some experiences are covered in the chapter 'Moving through the Bureaucratic Jungle'. ONGC also had its own bureaucratic culture which created problems with contractors, suppliers and environment around the areas of operation.

8. *Inappropriate Public Relations Work*

Any form of communication outside and within the organisation was non-existent except in the bureaucratic way, as prevalent in any governmental organisation. The Public Relations department was mostly busy in issuing advertisements and handouts on special occasions. The PR department did not have access to the information about the future and present except through annual reports and initiative of individuals in the PR department. There was a bureaucratic chain to get the information authenticated before it was released to the press.

A two way communication was established with people at various levels through DO letters non-agenda meetings at sites, to share challenges before the organisation and problems and constraints being faced at working level.

On 15th August and 26th January every year achievements and plans were covered in the address by the chairman, as also in annual reports. The DO letter dated 20th October 1981 issued immediately after taking over the charge as chairman, is at Appendix A. This was based on detailed analysis, of the challenges during my tenure of over 3 months as an officer on Special Duty. As an example another DO letter dated 16th August 1986 is at Appendix M to create the desired thrust for the years to come. The rank and file had responded with great enthusiasm, creating international recognition for ONGC.

9. Overpowering Influence of the Bureaucracy

In our system the bureaucracy gets threatened by successful professionals, as the events of the past and future would prove. After Independence, the development model in the country was based on the Soviet system, but for governance the British model was adopted. The generalists continued to govern while professional managers played only a subordinate role. This unfortunately continues to be the same even today and needs deft handling by the managers to succeed, else the bureaucrats dominate over the organisations whenever the organisational head is more pliable. This is the reason for the poor performance in ONGC from the early 90s when there was a change in top management, and the politicians and bureaucrats had their way. In the name of privatisation, the producing fields (Rava, Mukta, Pana, Tapti, and many other small fields) of ONGC were gifted away to the private sector who had no experience in the field of oil exploration and production. Privatisation for public good is understandable but not for obviously different reasons. All this exercise of privatisation was done in a hush-hush manner by downplaying the oil reserves. In fact, ONGC had the competence and technology to exploit the reserves in the best interest of the nation. **Even the marginal fields would have been put on production if ONGC was given international price, as given to the private players, after privatisation.**

11

Thrust on Oil Exploration

The history of oil exploration in India goes back to the nineteenth century, close to the first commercial discovery of oil by Col. Edwin Drake in August 1859 in Pennsylvania, USA. In India, the first commercial discovery was made in March 1867 at Makum in Assam not far from the first oil discovery in USA. (The history of oil till the formation of ONGC in 1956 and thereafter, is well covered in I. A. Farooqi's *The Story of ONGC.* S. N. Visvanath's *A Hundred Years of Oil* also covers the subject in depth.)

In pre-Independence India oil was discovered only in Assam and that part of Punjab which now is in Pakistan. It was only after the formation of ONGC in 1956, that the real thrust to the oil exploration activities in India was given, resulting in many oil/gas discoveries in onland Assam, Gujarat, Andhra Pradesh, Tamil Nadu and offshore, in Arabian Sea and Bay of Bengal. Inspired by the active support of visionaries like Malaviya and Nehru, oil exploration activities received a major thrust though the West remained completely indifferent and unsupportive. In this background of non-cooperation of the Western world, the former Soviet Union (USSR) however gave full support to train Indian scientists and engineers and provide experts and equipment on soft rupee payment basis. They also helped to establish two world class research and development (R&D) institutes of petroleum exploration and drilling technology. Initial discoveries in Gujarat and Assam and the discovery of the giant Bombay High oil field in the Arabian Sea are the results of the intimate support from the former Soviet Union.

Bombay High structure was delineated with the help of a Russian ship 'Academic Arkhangelisk' and the first well was drilled on the structure by the then acquired jack-up rig 'Sagar Samrat'. The oil field was discovered on 19 February 1974. The Bombay High oilfield structure is quite large by any standards and is located at about

160 km of the shores of Bombay. The depth of water at the drilling locations ranges from 75 to 90 m.

Soon after the discovery of oil field in 1974, the presiding incumbent of the organisation, B.S. Negi who had officiated as Chairman for around four years retired from service. The management under the chairmanship of B.S. Negi was conservative but highly committed to geo-scientific professional expertise which had resulted in the discovery of Bombay High field.

I had the pleasure of interacting with him a few months after I had assumed the chairmanship of ONGC. The results of the SWOT analysis of ONGC and the outline of long-term plan prepared by the task force headed by R.N. Misra, one of our bright senior executives, was explained to him. Negi was appreciative of the efforts and felt that the changes we were planning to bring about were radical but wished us luck. He however, expressed his reservations on the strategy followed immediately after his retirement for the development of Bombay High Field when for reasons not recorded, production from the field had acquired a new urgency and the field was put on production in great rush without proper reservoir assessment. He felt that a lot more delineation wells should have been drilled to fully appreciate the extent and potential of the field before putting it on such a scale of commercial production. The point was well taken, which indeed was also the viewpoint of many of our geoscientists. This approach was followed in tandem during my chairmanship that yielded rich dividends.

Indo-Soviet Integrated Exploration Project

Since the discovery of Bombay High and the induction of modern Western technology, the Soviets started feeling that they were being more and more marginalised. In the meanwhile two rounds of exploration bidding for offshore blocks were invited and awarded to foreign oil companies. The Soviets could not again take part in these bidding rounds because of their lack of offshore expertise. Therefore a justifiable fear complex was creeping in that the Soviet involvement in Indian oil exploration and development was getting more and more reduced. It was therefore decided through an Indo-Soviet co-operative agreement to make use of the Russian expertise onland, on terms at par with the Western oil companies operating offshore. The areas were to be chosen by mutual consent between ONGC and former USSR.

The Indo-Soviet integrated exploration project was signed during the transitional period between Mr Venugopal (former Chairman ONGC) and me and continued over the entire period of my tenure and even beyond. The areas allotted for this project were parts of Tripura, part of the Bengal basins covering Kolkata and further south-east, Ariyalur-Pondicherry depression in the Cauvery Basin and the Patan depression in the North Cambay basin.

ONGC Makes Its Presence Felt Internationally

The performance results of ONGC had reached far and wide. *Fortune* magazine had reported ONGC as the eleventh largest profit-making organisation outside USA. This brought focus on India from the governments in the developed world. ONGC was the favourite destination for dignitaries and delegations from abroad. A delegation from UK headed by their energy minister Sir Peter Morrison also visited ONGC.

A lot of interest was shown by international oil and service companies. A number of memoranda of understanding were signed. Schlumberger also set-up a joint research centre with ONGC (extract from *Petromin*, July 1989, Appendix L).

Initially involved only with upstream part (exploration and production) of the oil business, ONGC gradually spread its wings to the downstream business of oil in line with the vision projected in the 80s. It was only after the formation of ONGC in 1956, that the real thrust to the oil exploration activities in India was given, resulting in many oil/gas discoveries in onland Assam, Gujarat, Andhra Pradesh and Tamil Nadu, and offshore in the Arabian Sea and the Bay of Bengal. To appreciate fully the hazards and probabilistic nature of oil exploration, a few facts not generally understood by the policy makers and public in general need to be recorded.

Petroleum—Origin and Exploration

Petroleum was generated millions of years ago and occurs in the earth as gas, liquid, semi-solid, or in more than one of these states. Liquid petroleum which is called crude oil to distinguish it from refined oil, has an oily appearance and resembles somewhat the ordinary lubricating oil sold at petrol stations. Petroleum gas, commonly called natural gas to distinguish it from manufactured gas, consists of the lighter paraffin hydrocarbons of which the main constituent is methane gas (CH_4).

The origin of petroleum and the manner of its accumulation into pools (oil/gas fields) has not yet been satisfactorily answered by the geoscientists. The theory widely accepted is that oil and gas are formed by the decay of certain organic matter, under optimum conditions of temperature and pressure, which were deposited millions of years ago in the sedimentary rocks. Later it migrated and accumulated in rocks which are porous like sandstone or in the fractures and fissures of hard rocks like limestone, etc.

> ! It is akin to the life of army personnel moving into enemy territory where you have to provide for many contingencies... !

The search for oil starts with detailed ground surveys, by geological and geophysical methods, laboratory analysis of the rock samples collected in the field, and interpretation of the survey results keeping in view the geological domain where oil could have generated and accumulated. Till date, there is no proven method of direct detection of availability of oil in the underground. The actual presence of oil/gas can only be proved through drilling of exploratory wells which is the most expensive part of the oil exploration activity.

Oil exploration calls for determination and perseverance. In the Siberian Basin of the Soviet Russia, they had explored for almost 40 years and drilled 90 dry wells before they discovered huge quantities of oil and gas. Same is the experience in many oil-bearing basins of the world, including our own experience in Krishna-Godavari, Cauvery and Rajasthan basins.

Oil exploration is a hazardous activity. Those involved in this activity have to traverse and move through swamps, river valleys, dense jungles, rocky mountainous areas and offshore surveys with associated risks. All this calls for very determined leadership and intense training to face hazardous conditions, hostile environment and inclement weather. It is akin to the life of army personnel moving into enemy territory where you have to provide for many contingencies and surprises.

Exploration activity is highly risk prone-a sort of scientific gamble. It is therefore not without reason that Monte Carlo simulation and Delphi techniques are employed while assessing a virgin basin or a virgin field. Worldwide experience indicates that 49 per cent of the hydrocarbon resources are located in 49 giant fields while the remaining 51 per cent are in more than 30,000 medium and small fields. The discovery of giant fields is a cyclic phenomenon. Oil fields of the size of Bombay High are not discovered anywhere in the world in any time frame. It is closely related to technological breakthroughs and new concepts in the field of geoscience. Nature has endowed many countries with abundant, easy-to-explore hydrocarbon resources as in the Middle East, Saudi Arabia and Iran. Many countries at the same time have none or limited hydrocarbon resources such as Japan and Sweden.

After the discovery of the oil/gas field, a number of wells are drilled to estimate the quantity of available reserves. **Oil reserve estimates are at best speculative.** The initial estimated reserves generally, though not always, increase with time as more and more production data is made available and additional surveys are carried out. Likewise the oil reserves of Bombay High have increased manifold (almost three and half times) since initial estimates were made in 1975. So is also the experience with Ankleshwar and other oil fields in India and abroad. Unfortunately, this important fact was overlooked when the marginal fields of ONGC were literally gifted away to the private sector in the 90s on the pretext that they have meager reserves of hydrocarbon.

Oil Exploration During My Tenure

Interaction through brainstorming sessions with the geoscientists of all disciplines in various regions and institutes of petroleum exploration at Dehradun brought out the fact that not only is Bengal floating on oil, but also are Krishna-Godavari (Andhra Pradesh) and Cauvery (Tamil Nadu) basins. Major campaigns for surveys and drilling were launched which resulted in making these two regions (KG&C) major oil/gas provinces for subsequent exploitations both onshore/offshore. The number of rigs inducted in these basins between 1981–89 is a yardstick of the exploration effort. Appendices G and H detail the explorations done with the help of graphs and performance during the period 1981/82–1989/90.

The dramatic fall of oil prices in 1986 induced a reduction of more than 20 per cent exploration expenditure in the world, but it had an opposite effect on ONGC. Mid-term review of the Seventh Plan was carried out. As a result additional exploration effort was planned for Krishna-Godavari (K-G), Cauvery and Tripura, which firmly established these basins as commercial producing basins.

Some of these discoveries, considered commercially marginal at the prices then paid to ONGC, were nonetheless further delineated by resorting to Early Production System (EPS)—a concept that was first introduced and practised in K-G field with great advantage for development by modular development concept. This did not require the government's prior sanction and for the first time was constructed and implemented at the project level (K-G) by a very competent mechanical engineer of the project, Mr N.R. Pillai who later was the general manager.

Visits to Rajasthan, Assam and Cauvery basins brought to focus the difficult working conditions due to weather and terrain resulting in poor performance. Here too, like in Krishna-Godavari, air-conditioned bunkers and excellent catering support was provided resulting in improvement of performance. This resulted in increase of drilling and seismic crew performance by 300 per cent and 500 per cent respectively. Shift pattern like that in offshore—15 days on and 15 days off were introduced in these and some other difficult land areas. All these measures resulted in improvement in the work environment and the bottom-line of the organisation.

The philosophy of 'let a 1000 roses bloom' was adopted in encouraging exploration thinking and building up of geological models for the various basins and different parts of each basin. Calculated risk was undertaken in testing these geological models, built painstakingly on available data and flavoured by controlled imagination. As a result, Gandhar field in Gujarat was discovered in 1984. This is a multilayered stratigraphic trap type of accumulation with oil/gas/condensate in different layers, and hence complicated for a unified development. The field was put on EPS in 1986, in less than two years of finding it, to obtain additional production and reservoir data for planning further development. Today this is a giant field.

Geological models in the Cauvery Basin gave the results of sustained exploration drilling for over two decades. Kovilkalappal was discovered as a commercial producer (1985), not far from Pattukotai, where the first few structural wells were drilled in the basin and in Narimanam, on the flanks of Karikal High, where 10 wells had been drilled earlier without commercial success.

Ravva field in the shallow waters of Godavari offshore was discovered in 1988. This is an accumulation against a growth-fault, the first such discovery in India, otherwise prolific in the Niger

Delta. In order to delineate the trend of the sands of this multi-layered reservoir, the field was also put on EPS. It is a great pity that this field was donated away to a consortium, led by a little known operator, Command Petroleum of Australia, subsequently taken over by Cairn Energy of Scotland, both of whom had built in their entire fortune from Ravva, in addition to their other Indian and foreign partners.

Neelam, another giant in western offshore, was a classical structural discovery in 1987. The structure D-18 in western offshore was a successful discovery (1985) based on the geological model of 'mud-mounds', prolific elsewhere as in Siberia. The field was put on EPS in 1989. Pasarlaupudi (1987) was the biggest onshore discovery in Krishna-Godavari basin.

Besides, blocks earlier offered to private companies for exploration and surrendered by them without success, were re-opened after re-interpretation of data. Kutch (KD-1, 1984) and Godavari (GS-15, 1988), proved to be oil-bearing. It is a pity it took almost a decade and a half for GS-15 to be put on production.

The above are only a few instances to illustrate how a change in geological concept and improvement in technology had helped in adding low cost discoveries and reserves, higher than the targeted figure, during the Sixth and Seventh Plan periods. These goals, both for development of Bombay High and for diversion of resources on exploration were not set by anybody for the organisation. ONGC had set its own goals; these became a vision and the achievements of these goals became the mission. It gives immense satisfaction when these missions are achieved ahead of time and richer than planned. News items on exploration activities in Krishna-Godavari and Cauvery basins are appended below.

Krishna-Godavari Basin: ONGC Stepping Up Exploration

The Oil and Natural Gas Commission (ONGC) is giving a major thrust to its exploration activities in both onshore and offshore areas of Krishna-Godavari basin.

The ONGC Chairman, Col S.P. Wahi said on Saturday that the Commission had chalked out an action programme to drill series of wells in the Rajamundry belt.

Col Wahi, who had recently visited Andhra Pradesh, said the ONGC would deploy four more rigs to drill at least half-a-dozen wells in the belt. At present two rigs are being operated in the belt.

The Krishna-Godavari areas has also been included in the foreign oil companies for exploration and exploitation.

A seismic survey has found that the Krishna-Godavari area as 'very protective' and pregnant with hydrocarbon deposits.

In a Krishna-Godavari offshore structure the US rig Sedco-445 has completed the drilling of first well and there are indications of hydrocarbon deposits.

The first well was drilled to a depth of over 4,500 metres. The rig is now engaged in the drilling of a second well in the structure. ONGC proposed to drill at least another three wells to assess the commercial potential of the Krishna-Godavari offshore basin.

To step up the drilling operation in this offshore area, ONGC would deploy another floating rig. The Commission is developing the structure on a priority basis.

Source: *Financial Express*, 19 September 1982.

*****

Wahi Inaugurates Drilling at East Godavari

The chairman of the Oil and Natural Gas Commission (ONGC), Col S. P. Wahi, has inaugurated work on Oil well drilling operations at Kunavara, near Amalapura, in east Godavari district.

Speaking on the occasion yesterday, Col. Wahi, visualised a bright future for the areas where abundant oil resources were identified and said that efforts would be made to create more physical facilities in these areas.

Mr A. Farooqi, project manager outlined the technical aspect of the drilling programme.

Source: *Himachal Times*, 6 June 1983.

The vision of ONGC to go into the deeper waters has been validated by discoveries being made in the deep waters in the southern regions of the Krishna-Godavari basin.

ONGC Ignores World Bank Directive

The Oil and Natural Gas Commission (ONGC) has gone back to the G2 structure in the deep waters of the Godavari, where oil was struck in December 1982, disregarding World Bank's advice not to drill in structures beyond a depth of 300 metres.

ONGC sources say that the drilling of the assessment well G-2-4, which is in 500 metres deep waters will not have the World Bank loan cover and will have to be financed by ONGC itself.

The World Bank has advised the Commission not to venture into unchartered areas and confine its operation in the Godavari offshore to shallow waters. The bank felt that investment prudence did not call for such deep water drilling. Moreover, ONGC has not submitted to the bank any feasibility study regarding the technical adaptability, cost effectiveness of oil production from structures beyond 300 metres. Put differently, the Commission has not evaluated the financial viability of such deep-water drilling.

It is pointed out that the combination of high currents and high cyclonic storms make the Godavari offshore the most inhospitable area in the world for drilling. The bank's contention is that though there are claims of deep-water technologies by oil companies, they are still unproven. Their application is rendered more difficult in the Godavari which is much more inhospitable than the North Sea.

The *Economic Times* reported on March 25 last year that the World Bank had asked the Commission to confine its oil exploration programme under its loan cover to areas less than 300 metres depth. This was denied as 'nonsense' by Col. S. P. Wahi, chairman, ONGC. However, the World Bank norms incorporated in a confidential ONGC document read, 'The World Bank, Washington D.C., has advanced a loan of US $165.5 million to cover part of the exploration programme envisaged for 1982–85. The project area for the bank assistance is defined and demarcated to cover 17,000 sq. km. The seaward limit is restricted to 300 metres isobath.'

The World Bank-aided drilling programme for offshore projects envisages the drilling of 11 exploratory wells in the inner shelf area with an average depth of 4000 m and 5 wells in the outer shelf area with an average depth of 3500 m. 'The bank will bear 25 per cent of the foreign exchange expenditure on rig hire charges and 100 per cent foreign exchange expenditure as well as local expenditure for drilling materials and supplies', the document said.

The original schedule under the bank programme did not include deep water areas. But when the Commission insisted on drilling in G2 structure, World Bank reluctantly revised the schedule on the condition that all deep water structure drilling will be financed by the Commission on its own.

Source: *Economic Times*, 13 January 1984

*****

Exploration Activities To Be Accelerated

The Oil and Natural Gas Commission (ONGC) would accelerate its exploration in the Cauvery and Krishna-Godavari basins by adding more rigs soon Col. S P Wahi, ONGC chairman, said yesterday.

He told newsmen that the number of rigs engaged in drilling work in the Cauvery basin would go up from 4 to 10 and in Krishna-Godavari from 5 to 12. 'Our strategy is to reinforce the success obtained so far to achieve a breakthrough,' he said.

Mr Wahi, who along with other members of the Commission members is here for a review of the work of the southern regional business centre, said ONGC scientists were 'excited' about the high quality oil finds in Kovilkalppal and Narimanam in the Cauvery basin, where geological in-place oil reserves might be 20 or 30 million tonnes.

Col. Wahi said the additional rigs to be bought or charter-hired or transferred from other sites would be put into operation in the current plan period itself.

In the Cauvery basin a target of 0.5 million tonnes of oil to be produced by the end of the plan period had been fixed. A 100 crore rupee return on investment was expected.

In the Krishna-Godavari basin, ONGC intended to raise the gas potential to one million cubic metre per day by the end of the current plan and 3 million cubic metres per day by the eighth plan end. ONGC would take the gas by pipes and make it available to customers at specific points on 'trunk routes', he said.

Source: Northern Times, 10 May 1987.

Oil Exploration in the Eyes of Political Leaders

We in the ONGC had the good fortune of the full support of two visionary prime ministers (Indira Gandhi and Rajiv Gandhi) who because of their total commitment to the development of the country had realised the importance of oil as a political and economic weapon.

As mentioned earlier Indira Gandhi was very well informed about oil and the strategies being followed by other countries to manage this vital depletable source of energy. She had a deep understanding of the probabilistic nature of oil exploration and had known the strategy being followed by countries like USA. It was this realisation that made her ask me besides other searching questions whether we should exploit all the oil fields as and when we discover them. This question is even more relevant today from the point of energy security. She had rightly appreciated the strategic importance of oil to be retained under the control of public sector. This strategy was further reinforced by her son, Rajiv Gandhi.

Rajiv Gandhi, apart from his vision and implementation of other development activities, took very keen interest in energy security particularly self reliance in oil. On his advice I had presented before the whole cabinet committee on economic affairs the oil scenario and future possibilities. During the presentation which continued for an hour and a half, he interacted actively on the subject and suggested very important leads to act on. He also conveyed his appreciation of ONGC's performance, a rare tribute from the PM, which is presented below.

Prime Minister

Message

The Oil and Natural Gas Commission has been carrying out a very vital task for the nation. Any nation today is judged by the amount of energy it consumes. India during these past few years has increased its energy consumption tremendously. The ONGC has been largely responsible for the energy that we have been using. The ONGC has also demonstrated that the public sector can be efficient.

The public sector can compete with the best anywhere on the globe and I would like to congratulate the ONGC for its performance.

Source: 24 February 1989, New Delhi.

Rajiv Gandhi continued to interact with me personally even after my retirement on many issues in the interest of the country. In one of the meetings with Rajiv Gandhi I had once suggested the purchase of oil service companies in US as these were available for 'a song' due to the downturn in oil exploration activities in the US in those days. A very eminent economist who happened to be present at that time, however, put 'cold water' on the proposal by his remark, 'How can we spread our resources so thin?'

Visionary Changes in Oil Exploration

The new thoughts/initiatives during my tenure were projected in the statements of the chairman in ONGC's annual reports, apart from the DO letters from the chairman, as and when a new thrust was required to be given. This enabled the employees to get the information about the common objectives and the direction for the future, creating the desired motivation and enthusiasm for everyone to strive harder. A very close feedback system was developed to correct any error and to ensure smooth transformation.

ONGC during the period 1981/89 had achieved phenomenal growth and international recognition. On every opportunity stress was laid on oil exploration activities (see Appendix D&N). The credit goes to the innovative and creative minds of its people. We all worked together for the common objectives. A culture of collaboration prevailed. ONGC became a role model for the corporate world.

Some of the proposals which were initiated ahead of times (integration of oil sector, exploiting hydrocarbon resources in deeper waters, vertical integration of ONGC, acquisition of exploration and production acreage abroad, diversification into power sector, use of gas for automobiles, In-Situ coal gasification, National Gas Grid, etc.) were set in the process of realisation. In one of his letters dated 31 October 2005, Mr Subir Raha, the then chairman of ONGC had written, "I find it a privilege to bring a few of your dreams to reality."

New Exploration Concept—Common Basin Approach

As mentioned earlier, it was strange that in an organisation whose core business is exploration did not have member-exploration in the Commission for almost 10 years. There was lack of team work and synergy between the various disciplines (geology, geophysics, chemistry, laboratories, etc.). The organisation structure was based on offshore and onshore. The offshore and onshore divide had resulted in missing vital links during exploration in areas where there was extension of oil fields from onshore to offshore as in Krishna-Godavari and Cauvery basins. Activities were being handled from Mumbai for offshore and Chennai for onshore. One group was looking west and the other east and missing the important leads. During one of the informal meetings with geoscientists, it was brought out that oil formation underground does not differentiate between land and sea. I recall Mr Y.B. Sinha, a bright young geoscientist who later became director (exploration), was involved along with others with the concept of Common Basin approach, which laid the foundation for excellent results in subsequent years to discover more oil/gas fields. This concept gave rich dividends in K-G and Cauvery basins. But it is a pity that highly prospective finds in the Krishna-Godavari offshore areas were pawned off to private operators and the benefits of conceptualising and discovering the same could not be tasted by ONGC. This action of the government of 'Robbing Peter and Paying Paul' has demoralised the ONGCians, who through their sweat and blood had discovered the oil fields. The fruits of their efforts was gifted away. In consequence, the oil production from 1989/90 has been on the decline instead of being on the increase as it was till then.

Research and Development

During interactions with young scientists and engineers in existing R&D institutes, it was decided to make R&D institutes as notional profit centres so that majority of the work in these institutes could be sponsored by the regions with live technological/scientific problems. Another important factor recognised very early at the stage of formulation of our new strategies was that the challenges that the Commission had taken upon itself for accelerating exploration and production activities could not be fulfilled without research and development being designated as a 'key result area'. Accordingly R&D activities were intensified in the three existing institutes of the Commission. Research and development received very special attention and many innovative actions were taken to achieve results. A Research and Development Coordinating Committee was formed to ensure actions. Updating technology and research on field problems received greater attention. The three institutes—petroleum exploration, drilling technology and reservoir studies got increasingly involved with field problems. Maximising recovery of oil (recovery factor) from the existing oil fields, particularly offshore, became a serious matter of research. The institute of drilling technology was also entrusted with the task of training new entrants and organising refresher courses for in-service drillers. The basic objective was to update technological information and skills and to develop every single employee as to enable him to give his best. In consonance with these ideas two more institutes—Institute of Engineering and Ocean Technology and Institute of Production were established. Having adopted 'environment protection' as the fifth corporate objective, an institute of petroleum safety and environment management was also set up at Goa.

The impetus in R&D operations related in-house activities and farming out basic research to the universities and other scientific organisations enabled R&D institutes to provide intimate support to solve operational problems of the regions as a matter of great urgency. This enabled optimisation of resources in the operating regions. Research and Development institutes were able to further accelerate consultancy work for oil companies in other countries. In the process a few technical patents were also registered by ONGC's scientists and engineers.

Linkages with Universities and Scientific Organisations

It was decided that most of the basic research work be sponsored to laboratories under the Council of Scientific and Industrial Research (CSIR), universities and other such bodies established in the country. With this in view, a two-day conference was held in November 1985 with universities and other scientific organisations to establish linkages for farming out basic research from ONGC R&D institutes to universities and other technical institutes and also to discuss and suggest ways and means to arrive at a meaningful solution of a more effective collaborative effort between the oil industry and others.

*****

In conclusion I would like to reiterate what I mentioned earlier: the game of oil exploration is a scientific gamble; its input is deterministic but output is probabilistic. With more and more data being available in recent time, through additional surveys the prognosticated hydrocarbon resource base is on the increase. More than 25 billion tonnes of hydrocarbon prognosticated resources have been projected in 2006, which have still to be converted into geological inplace reserves through exploration. No hydrocarbon discovery has yet been made in 20 virgin basins. These are high risk and possibly high reward areas. Special strategy may need to be formulated to induce oil companies to explore in these areas and special funds need to be created for the purpose. There is also a lot of potential in the old fields. The basement in Bombay High Fields for example contains a lot of oil in the fractures calling for R&D efforts to extract oil from the fractured basement. Efforts should continue with more vigour so that ONGC gets access to oil not only in basins in India, but also abroad.

12

Business Is Like a Battle Front

How true is the Chinese saying that business is like a battle front (*Art of War* by Sun Tzu, the Chinese philosopher)? Some of the tactics, principles and strategies and traditions followed in the army, which have stood the test of time are equally applicable to a successful business.

- As the supreme commander develops a team of commanders in the army so also is a management team developed in business.
- As the strategic plan is continuously updated in the army taking into account enemy strategies and tactics, so also in business technology, strategic operational plans need to be continuously updated.
- Surveillance of likely enemy activities is an important function in the army. In business too, plans of competitors are to be continuously monitored.
- In army continuous training and retraining for upgradation of knowledge and educational qualifications is an important activity, as in highly scientific and technologically oriented business.
- Constant communication at all levels about challenges and tactics is shared even at the lowest level of troops at daily evening roll calls. Similarly, open communication channels with different methods are important for business.
- Army works mostly on matrix organisational structure. Everyone is clear of the objective—defend the safety and integrity of the country and win battles and wars to survive. In business, it is also the survival of the fittest. The structure ensures interdependence between various arms and services. Integrated

task forces are created for special tasks. The same tactics have to be followed for a successful business.

- The growth of officers is watched carefully in the army and takes place in two streams of staff and command. Those selected for command have to have extraordinary powers of taking initiatives and making timely decisions. They have to be role models for the troops and should have the ability to inspire and influence them to move enthusiastically and willingly to achieve common objectives. Similar streams of command and staff are also initiated in a successful business organisation.
- There is harmony in working through traditions in different units and formations of the army. The uniting factor is the name of the unit or formation. Youngsters joining the units are well groomed about the traditions and culture. For effective business management such practices are equally important.
- Reinforcing success is a very important principle to win. Review and consolidation for the next step, in situations where results are not up to expectations is a continuous process in the army. The same is true in business.
- Adherence to the policy of 3Fs (friendliness, fairness and firmness) in the management of men in the army is a golden principle; nothing can be truer for success in business than following these ideals.
- Every leadership decision takes into account three factors — morale, motivation and available resources of the troop. This mantra is equally applicable to business to get total commitment of the people.
- In operational areas, empowerment initiates one to attain the target. Empowerment has similar motivating results in business.
- More importantly a lot of love, affection and prestige are given at all levels. No work is considered too small. Everyone puts the hand to the task—whether sports or other activities. Similarly in business too, leaders have to lead from the front.
- Management of external environment, i.e., politicians, bureaucrats and agencies who can supply information and give assistance, is equally true and important for successful business.

Action Plan and Strategy for Change

After joining ONGC, I interacted with a wide cross-section of employees in every area of operation and sought their views about their problems and asked for their suggestions to solve them as well as to augment the growth of the organisation. A rapport was established both at the formal and informal levels and excellent suggestions were received. Studies of selected multinational oil companies about their organisation and functioning were also undertaken. As a result some key action plans, as listed below, emerged which required immediate attention.

- Long-term strategic management plan to ensure continuous oil exploration effort to convert large prognosticated hydro-carbon resource into geological inplace reserves.
- Organisational Restructuring – offshore and onshore barrier.
- Synergy in exploration of different disciplines.
- Management services groups for external and internal communications, environment scanning and management of change.
- Human resource participative management and management by wandering, team building, interdependency at corporate level, effective advocacy of the 3M mantra (motivation, morale and money being the bottom-line), and the 3F mantra (friendly, fair and firm).
- BPO (Business Process Outsourcing) of the low-end technologies or operations which are not core functions, like Seismic Shot-hole drilling, transportation of both men and equipment, cleaning and house keeping, fabrication of spare-parts, etc. to improve 'teeth to tail ratio'.
- Massive indigenisation effort so as to arrest the drain of foreign exchange and promote Indian industry for the manufacture of drilling rigs and components, offshore supply vessels, offshore well platforms and production platforms, etc.
- Training and development for managerial and leadership roles.
- Espirit de Corps and unity – collaborative approach.
- Growth of the individual identified with the growth of the organisation.
- Social welfare and invisible merger with the environment of work and operational areas – societal obligations.

As mentioned earlier, challenges ahead were colossal but these were exciting and loaded with possibilities for success. I drew heavily on army experience in strategic planning, principles of war, operational tactics and management of human resource (the main factor for success in any activity). I had rubbed shoulders with outstanding corporate leaders, both senior and junior, in my earlier assignments. This experience was of great help in taking the following remedial actions.

> ❗ It is not enough to do good work but people should know about it, to establish credibility and name for the organisation. ❗

Management of Change

I had made extensive notes of the changes required to transform the organisation to the best international standards. A deliberate plan was worked out and shared with colleagues at various levels. The changes were radical. However, while implementing these changes, we tried to be conscious of the fact that change creates a feeling of insecurity in people's minds. Hence change needed to be managed through proper deliberate strategies and tactics. A DO letter (see Appendix A) covering the future policy directions was issued and widely circulated. This prepared the people for the future and enabled them to undertake critical analysis. They did give very useful inputs during my interaction on periodic visits. DO letters were issued periodically. [See Appendix M)

Building Up Awareness – Communication

It is not enough to do good work but people should know about it, to establish credibility and name for the organisation. The information to our own people was essential for boosting their morale and also for building up their confidence and making them aware of the role being played by ONGC in the vital core sector of energy. It was decided to set up a management services group at the corporate level as well as with each member. They were made responsible for monitoring the performance of the Regions as well as Management Audit with a view to reduce costs. They were also entrusted with the task of scanning the economic and technological environment and preparing the status papers on various aspects of energy to work out future options and opportunities for ONGC. In effect they were to act as the 'eyes and ears' of the ONGC in terms of future planning and policy initiatives.

In view of the importance of the energy sector in which ONGC was operating, the media was always looking for information. This opportunity was effectively used by issuing handouts to the press which enabled the ONGC personnel, spread all over the country, in getting live information. This also established a good rapport with the press as well as with other media of communication. The costs on sponsored advertisements were drastically reduced. An extract below explains the changed position.

A PR Case Study, Oil's Well at ONGC

The Oil & Natural Gas Commission (ONGC) has a continuing love affair with the Indian public, as the public are the sole shareholders of this Corporation. There is a new thrust to Public Relations under the new Chairman, Col S.P. Wahi, who has re-oriented the Public Relations function and made it report directly to him, locating it at the Dehradun headquarters with the other PR units strategically located at Bombay, Baroda, Calcutta, Delhi, Sibsagar (Assam) responding to local needs under policy guidance from HQ.

The unique feature of ONGC is that there is a story every day, there is so much happening and this is eagerly reported by the press, so that ONGC has not needed to release the large-space advertising that it initially used (1975-78) because through the editorial coverage it now regularly gets its best publicity.

National and international significance of ONGC operations are well known, but each of these editorial mentions/highlights how rapidly results are being achieved. The Public Relations department has to make sure that each item is lively and not drab.

Source: *Promotion,* August/September 1982

M. L. Kaul, head of PR department, reoriented the work culture to this new strategy of management through his professional skills. A very aggressive campaign for communication within the ONGC was launched through its in-house magazine *Reporter.*

Constant communication through periodic demi-official letters to people in ONGC brought out challenges facing the organisation. During visits to the worksites/projects I met people freely and listened to their problems and suggestions for improvement in performance. This was in a way the most effective two-way communication.

Attitude surveys were conducted particularly among the younger colleagues through structured questionnaire to tap their innovative minds and get their feedback about the work culture. Thereafter a number of executives were met on receipt of their replies, to decide on changes required to improve organisational performance.

> ❗ Human resource has unlimited creative and innovative thinking capacity. ❗

A very close interaction was maintained at all levels with the local people and administration in areas of operation. Necessary assistance was provided to the schools and industry, particularly small-scale industries. These enabled to develop very harmonious relations with the environment with the result that even during worst political agitations in some states, the operations of ONGC were not disturbed. This issue is elaborated further in the chapter—'Management of Environment'.

Human Resource Management—the Fundamentals

The political, economic and technological changes in today's globalised world have its own impact in business. Therefore, the way we think, operate and do business in present times has changed. Disintegration of USSR, integration of Europe and rejuvenation of Southeast Asian countries, capital market fluctuations and impact of oil price volatility, bring to the fore new management thoughts and strategies. Management gurus/academicians continue to come out with management jargons reacting to changing situations, study of a few corporate houses or technological developments. These do not always stand the test of time, as in most cases they relate to specific situations, while business environment is in a state of continuous change.

A few fundamentals of management particularly those related to human resource have neither changed nor are likely to change with time as ultimately results are achieved by the people. Proper management of this resource is the key to success particularly during difficult business environment and crisis situations. Human resource has unlimited creative and innovative thinking capacity. Emotional intelligence of leaders helps to create team work, espirit de corps and commitment to achieve objectives. It raises their morale, self-esteem, mutual respect, prestige, dignity and trust all of which are so essential to get the best out of the human resource, and to tap their creative and innovative minds.

Krishna-Godavari and Rajasthan are classic examples of how the scales got turned to the advantage of the organisation when sagging morale of the drill crew was boosted up. Drilling in these areas had been going on for some time but all the wells suffered stuck-up or some other problem with the result that while the areas were known to have hydrocarbon, no well could come to the testing stage, and consequently no oil or gas could be produced on surface. One of the reasons behind this was the low morale of the drilling crew who, because of their shift pattern of a week's work and 1 1/2 days off, had to remain away from their families (crew was mainly from Kerala and Himachal Pradesh) for a long stretch of time in sheds with poor basic facilities and in extremely hot climate. Once they were provided air-conditioned bunk houses close to the drill site with messing facilities on the scale as in offshore, and shift pattern akin to offshore practices with both way fares paid to their hometown on their offs, things dramatically changed. Rajole-1 was the first well that flowed gas at tremendous pressure and this by itself rejuvenated the spirit of the entire crew and staff of the project with the result that thereafter there was no looking back, and one after another fields were discovered in that area with no serious drilling complications. This further proves my point that if one looks after the welfare of the employees, the latter will give the best in return—there cannot be a better mantra in management.

Strategy for Growth

To lift the morale of the people it was important to prepare a long-term strategic plan involving the employees so that they could see their own growth with the growth of the organisation. Through formal and informal (tea cup) meetings, the innovative and creative minds of the people were tapped. Over a period of time a participative culture was developed based on trust, mutual respect, mistake tolerance, but more importantly, healthy dissent. The sensitivities of the people and their dreams of personal development were noted for policy initiatives. They were insulated from bureaucratic needling and were made proud for their contribution to the organisation, which they rightly deserved.

Team Building Corporate Level

A vision to be ahead of times was developed. It was essential to have an integrated cohesive team at the corporate level. A number of non agenda meetings were held with the members of the Commission based on the feedback received from the rank and file through formal and informal channels.

> A vision to be ahead of times was developed. It was essential to have an integrated cohesive team at the corporate level.

Chairman and members visited the operating regions together to review the performance and provide the support on the spot. This avoided movement of files. For promotion to senior positions, the candidates were interviewed by the chairman and all members of the Commission as part of the selection board.

A management committee of the members headed by the seniormost member had been formed to resolve issues, affecting the progress. The new organisational structure had created interdependence between the members to meet targets set for the six regions. Each region was the responsibility of one specified member including member (HR) and member (finance).

In addition, one to one meetings were held with a large number of senior executives. This helped to create a common wavelength and common objectives to exploit the hydrocarbon resource base. Attitudinal surveys were conducted, up to the middle management level, to get the view on the environment and work culture to improve their performance. A large number of executives were met, particularly those who had made suggestions for improvement. Actions were taken to make changes where required. This helped to improve the work environment and create a sense of belonging.

New Planning Model—Twenty-Year Perspective Plan

India has a very large hydrocarbon resource base of 28 billion tonnes (in 1981 the estimate was about 15 billion tonnes). These prognosticated resources are spread over 26 sedimentary basins covering an area of over 3.14 million sq. km. Only 6.5 billion tonnes of geological in-place reserves have been established by exploration. Earlier planning based on a five-year cycle resulted in the highs and lows in exploration activities. A long-term Twenty Year Perspective Plan was prepared in-house to ensure continuous acceleration in exploration activities and exploitation of a huge prognosticated hydrocarbon resource base. It was perhaps for the first time in the history of corporate sector that a 20-year long-term plan was prepared and implemented. Based on the 20-year plan, a 10-year accelerated plan was prepared and presented to the important policy makers (see extract below). The Seventh Plan outlay was increased by 50 per cent without initiation of any paper work. It was a novel initiative.

Novel Method of Getting Commitment of Policy Makers

The Oil and Natural Gas commission has sought a substantial increase in the financial allocation from the Union government for implementing its accelerated plans of exploration and exploitation, it is learnt.

To apprise the top brass of the vast oil potential in the country and the plan to achieve sizeable increases in production, the ONGC chief, Col S.P. Wahi recently made a presentation of the commissions' ten year plan. Among those who attended the audio visual presentation by Col S.P. Wahi included the Cabinet Secretary Mr Krishnaswami Rao Sahib, the principal Secretary to the Prime Minister, Mr P.C. Alexander, Mr Mohd Fazal, Prof. M.G.K. Menon; Dr C.H. Hanumantha Rao, Dr Manmohan Singh (Member Planning Commission), Mr V.B. Eswaran, Secretary Finance, Mr G.C. Baveja, Secretary (Expenditure), Mr T.R. Satish Chandran, Secretary, Deptt of Energy, Mr Lovraj Kumar, Secretary Petroleum and Dr S. Varadarajan, Secretary, Deptt of Science and Technology. A number of other senior officials from the Prime Minister's secretariat, other economic ministries and members of ONGC were also present.

The novel method of apprising the policy makers of the problems and prospect with the help of charts, graphs and pictures was appreciated by all those present .Following this presentation, ONGC circles believe that it would be easier to obtain sanctions for additional finances.

Source: *Times of India,* 18 June 1982

Long-term planning remained the strategy for growth and maintained the motivation for the employees for their own growth. **Everyone within ONGC was made to believe through regular interaction that long-term planning has to be the culture, and long-term goals should not be sacrificed for short-term gains.**

Tacit Approval of the Government

As a well thought of strategy, important issues were covered in the statements of the chairman, reproduced in the annual reports of ONGC, to get the tacit approval of the government, as all ONGC reports had to be presented in the parliament. This strategy was used effectively to get over the handicap of the ONGC Act in getting approvals through the slow-moving bureaucratic machinery. They never caught on this tacit strategy as the performance of ONGC overshadowed their expectations.

Through this strategy only those cases involving foreign exchange were sent to the government, and the rest were approved within the Commission. I may quote, as an instance, how certain modifications to Travel Allowance/Daily Allowance rules were introduced in spite of negative delegation of powers in the ONGC Act. Executives were allowed to stay in hotels of different categories depending on their level of designation, but for meals they had to depend on their daily allowance of Rs 85. Any increase in the allowances through the bureaucracy would not have been enough to have three meals in the hotel of stay. To circumvent, administrative instructions were issued that DA will be proportionately deducted depending on the number of meals taken in the hotel. We acted within the rules, as the amount of DA was not touched for any increase. They could now live with dignity while on temporary duty and avoid obligations of friends or to be on the look out for cheap *dhabas,* etc. This reflected on the improvement of individual performance, as they could spend more time on work instead of looking for food within the authorised DA of the princely sum of Rs 85.

Mission and Objectives

Simultaneously, over a period of time, a mission and objectives were laid down to provide a vision to the people and develop the strategy of working together.

The mission was to stimulate, continue and accelerate efforts to develop and maximise the contribution of the energy sector to the economy of the country. The objectives were outlined as— self-reliance in oil; self-reliance in technology; assisting conservation of oil, more efficient use of energy and development of alternate sources of energy; environment protection; and promoting indigenous effort in oil-related equipment and services.

Collaborative Approach—Participative Management

Apart from strategic management, to develop core business, human resource management received the highest priority. Members of the Commission set the pace as an united group, to bring about a culture of friendliness, fairness and firmness. This resulted in total commitment of the rank and file to the mission and objectives of the Commission. There was total transparency in working. The policy decisions were taken through participation of all concerned. The three mantra (morale and motivation of employees and money the bottom-line of the organisation) prevailed in decision-making.

Challenges before the Commission were communicated widely to get the feedback. Innovative and creative thinking groups were established to tap the innovative and creative minds of the people. Excellent ideas were received which resulted in actions. This helped the Commission to be ahead of times. All these were recorded in the annual reports of ONGC. They reflect the combined wisdom of my colleagues at all levels, on the various strategies and tactics to improve the performance of ONGC, its people and corporate responsibility to the society.

Adherence to the 3Ms

I had learnt very early in my working life in the army and later with two stalwarts (Mr Mantosh Sondhi and Dr. V. Krishnamurty) that the 3M mantra needed to be adhered to before any major managerial decision was taken. All the three Ms are interrelated.

Long-standing unresolved issues, had resulted in the low morale/motivation of the employees apart from the fact that a culture of confrontation instead of collaboration had got created. Even the officers' association had in the past taken a confrontational attitude with the top management. How can one afford to be callous to the legitimate needs of the people particularly when the organisation is generating surplus/profits with the efforts of the employees?

Through a campaign most of the pending issues were resolved and communicated throughout the organisation through administrative channels and periodic DO letters from me. An extract from the DO letter No.6/26/CM dated 25 May 1982 is presented below.

You will please recall that during these recent months we have been in touch with each other either directly or through our representatives. These meetings have been very valuable for the reason that they provided an excellent opportunity of understanding first hand the several difficulties being faced by our workers-colleagues in various work centres of the ONGC. The meeting also provided several suggestions for possible action and the Commission has greatly

> ... culture of confrontation instead of collaboration had got created.

benefited by your highly constructive ideas and suggestions. I thought it might be useful to review the several matters we have discussed together and to take stock of what we have been able to achieve and what should be our plan of action for the immediate future. I have written all this with a view to assuring you how deeply the Members of the Commission as well as I are concerned over the problems relating to our worker-colleagues and how it will be our endeavour to continue to make more and more efforts to bring about further improvements for the well-being of all our worker-colleagues in future as well. This you will readily appreciate will also depend on the increase in our financial resources that can result directly only out of the increase in our own efficiency and effectiveness. While we are no doubt concerned at all times about the ways and means of improving the quality of the life of our worker-colleagues, it goes without saying that all of us should be equally concerned with improving the quality of our operations in every aspect of the working of the Commission. I would like to re-emphasise that the Commission's personnel objective is to bring about a participative style of management to ensure reasonable working conditions, job satisfaction, emoluments and career advancement commensurate with performance as well as Commission's growth, respect for each individual, regard for his well being as well as the well being of his family and goodwill and spirit of friendship, understanding and team work amongst all the employees of the Commission. We want to make the best utilization of manpower and material resources for the good of our Nation in the best traditions of our patriotism and love for our fellow citizens. The Commission has brilliant future ahead. You are aware of the initiative taken in drawing up a ten year –Accelerated Growth Plan, for accelerated exploration as well as exploitation. This plan is intended to provide more than ample growth opportunities for all our worker-colleagues and I have no doubt you will join me in planning and executing for this bright and fruitful future ahead of us. I would finally add that we should all work with mutual love and confidence.

The important issues which got covered were the introduction of incentive scheme, promotion/upgradation of Class III and IV employees, employment assistance to dependents of deceased employees, enhancement of the drilling allowance, liberalisation of leave fare assistance and transfer grant, enhancement of conveyance maintenance expenditure, reimbursement of local travel charges while on tour, remote locality allowance to employees of Tripura Project, grant of additional cash allowance for arduous duties, enhancement of the amount of various advances, removal of pay fixation anomalies, simplification of procedure for grant of annual increments, settlement of claims of ex-deceased employees, setting up of central schools, liberalisation of merit scholarship scheme, housing facilities, decentralisation of work of sanctioning house building advances, allotment of gas connections, promotion of sports activities, formulation of ten years welfare plan, etc. A number of other benefits for the welfare of employees, which were under consideration, such as training facilities were also intimated in the same DO letter.

Culture of Mutual Respect and Trust

In line with the policy of the 3Fs (friendliness, fairness and firmness), indiscipline of any type was handled firmly. A group of people who entered the office building during lunch break were suspended. The suspension order was revoked only when their MP (Member of Parliament) accompanying them gave a written apology. One of the presidents of a union had advised his constituents not to put any demands as ONGC management was always ahead of the demands from the union. Along with a Twenty-Year Perspective Plan, a Five-Year Welfare Plan with the involvement of the unions was also prepared. The working was transparent. Doors of the management were always open after working hours to meet anyone who had some problem or issues to share.

Self-Reliance of Equipment and Services

A very strong industrial base had been established after Independence but due to lack of support to the indigenous industry by the corporate sector, equipment, offshore structures and high tech services continued to be imported. Having had the experience in the development of indigenous industry during my service with steel, cement, heavy electrical and engineering, I was confident that the indigenous industry would meet the requirement of ONGC for equipment, offshore structures and services. It was decided to set one of the ONGC objectives as 'Promoting Indigenous Effort in Oil Related Equipment and Services'.

> I was confident that the indigenous industry would meet the requirement of ONGC for equipment...

Long-term plan was widely disseminated within the ONGC, government and industry. The industry could appreciate the requirement of equipment, materials and services and could take long-term investment decisions. This brought in literally an industrial revolution. A large number of enterprises got established in the private and public sectors to meet the projected demand of ONGC and associated activities. The government also agreed to provide financial incentive for indigenous supplies and services.

The number of offshore supply vessels almost trebled in the period 1981 to 1989. In 1981, offshore supply vessels were 100 per cent foreign flag, whereas in 1989, they were 100 per cent Indian. In 1989–90, there were 67 offshore supply vessels, of which 33 were owned by ONGC and the rest by Indian private companies, including 10 by Shipping Corporation of India. In the same period, manufacture of drill ships, jack up rigs for offshore drilling, offshore structures and other highly sophisticated equipment for oil industry got manufactured indigenously. In the process, technology transfer to the Indian companies through joint ventures with foreign companies took place, creating possibilities for export of equipment, materials and services to other developing countries. Drill-rigs and ships with Indian flag are today operated by these private companies profitably, in many foreign countries.

'Indigenous Development'

Dependence on Foreign Consultants

A very strong culture of dependence on foreign contractors for the development of Bombay High Field and manufacture of associated offshore platforms, pipelines and other activities had suppressed the development of indigenous capabilities within the public and private sector. It was also necessary to encourage our own geoscientists for steer heading exploration activities. A balancing strategy was worked out whereby the guidance of foreign consultants like CFP was retained for sometime more with the association of ONGC geoscientists and engineers, to ensure the transfer of technological know-how at the same time. This issue has been dealt in greater detail in the chapter on 'Bombay High Oil Field Canards'.

The consultants never accept responsibility, but only defuse accountability. Basically consultants only use internal information and expertise and package the results well to impress others. On case to case basis, one sought advice from outside experts, but the accountability had to rest with the organisation.

The CFP contract for Bombay High Development was not extended beyond 1985 in the then existing form. Instead, after prolonged negotiations a contract was signed mentioning that ONGC would seek CFP's advice on specific projects, referred to them by ONGC in any field or topic, offshore or onshore, and that CFP would engage a few of ONGC's experts also on payment by CFP at international rates in any of their projects in India or elsewhere. This was some time in 1986 and CFP was naturally unhappy over their loss of control over Bombay High.

In March 1989, project was identified for cooperation between ONGC and CFP on the reservoir evaluation of Bombay High. In June 1989 a draft report was submitted by CFP on this study. The report stated, inter alia, that because of declining reservoir pressure, increasing trend of Gas-Oil-Ratio (GOR) and imbalance between production and injection, there will be a reduction in the final recovery of oil and fear of loss of oil to the large gas-cap area in both Bombay High South and Bombay High North. The draft report was considered by ONGC highly vitiated, perhaps deliberately motivated to ring false alarm bells. The report was considered unacceptable since CFP had not taken into account many critical technical parameters like the role of critical gas saturations and the pressure at which solution gas may be released in the reservoir itself to cause any irredeemable damage.

I therefore deputed Mr P. K. Chandra and Dr. S. Ramanathan to Paris to discuss our observations in the report with CFP. After detailed discussions, CFP agreed with our viewpoint and withdrew the draft report.

Organisational Rejuvenation – New Pride

The SWOT analysis by the ONGC team had brought out the neglect of exploration activities. This called for drastic measures to 'infuse blood' into the system for energising the people to meet the future challenges and opportunities brought out in the Twenty Year Perspective Plan to exploit huge unexplored hydrocarbon resource base. During their interaction with people in the Regions, and while preparing the SWOT analysis and the long-term perspective plan, the ONGC team headed by R.N. Misra, deputy general manager (he became regional director later), projected the bright future of ONGC's growth and created the desired wave of enthusiasm among the employees. Slowly but surely a new pride and culture of development took place. Everyone accepted the slogan: 'growth with stability and continuous improvement in productivity', for which internal strengths were marshalled.

Subservient Culture

The work of member (personnel) was being handled by member (finance). He was a brilliant finance professional, but was overloaded with two vital functions. Prior to him this function was being handled by an eminent IAS officer. Member-onshore and member-offshore had so many diverse technical disciplines and wide regional responsibilities that they had no time to manage professional development of the technical people. As mentioned earlier there was no member-exploration for nearly ten years, therefore professional development of geoscientists in an organised manner was non-existent. The presence of foreign consultants and contractors had suppressed the motivation to achieve professional excellence. The inherent potential of the executives in general was very high but they were afraid to act for fear of failure. A 'Yes' culture, stifled initiative and compelled one to respect rules and procedures rather than focusing on performance, had crept in. Tolerance of mistakes was very low. **Creative and innovative minds were not encouraged to bloom. A culture subservient to the consultants and contractors had developed.** This came out very clear even at a very senior level during the management of the Sagar Vikas Blow-Out (refer chapter on Sagar Vikas Blow-Out). It was decided to induct Mr R.Srinivasan as member-personnel. He was an outstanding professional who was my colleague in BHEL. He was of great help in transforming the organisation and making the people believe that the management cared for them during their service and after retirement.

Acting Beyond Laid Down Powers

ONGC's hands used to remain tied by the Act of Parliament which had numerous amendments brought about by bureaucrats without being accountable for the effect on performance. They had even amended powers of the chairman so that he could not take any decision without its clearance from the Commission and the government. This meant that none had the powers to implement a change in the organisation or take a decision if the Act was to be followed. I made a mental decision to overlook both the R&P regulations and the ONGC Act without disclosing my mind or making any formal directions. I was however aware that to acquire the power to act, one has to be backed with outstanding performance. To take decisions at the right time is the key to achieve excellence in performance amidst a changing business environment.

> To take decisions at the right time is the key to achieve excellence in performance amidst a changing business environment.

A related area of concern was the rapid increase in the activities of the ONGC, both in exploration and development in onshore and offshore. This required not only induction of large manpower at the junior executive level but also accelerated promotion of the existing executives to man appropriate positions in the hierarchy. Thus, the existing R&P regulations were relaxed on need basis and accelerated promotions were given. This practice was reviewed periodically to meet organisational needs for efficiency and balance. The above decision to overlook R&P regulations and ONGC Act were triggered by a few incidents, in addition to the stagnation in the growth of people and the large unexplored prognosticated hydrocarbon resource base, calling for accelerated activity beyond the comprehension of the existing status quo culture.

A widow had come from Ahmedabad to see the chairman to recover the dues of her bereaved husband. She had left her two young children at home with a neighbour and had come prepared to stay for ten days. When the matter was brought to my notice, I advised the executive from the personnel department to ignore the provisions laid down in the ONGC Act and other regulations, and act personally on such personal matters, and not impersonally. He was instructed to make arrangements to pay for the lady's transportation and send her dues to her residential address within 21 days.

On another occasion, member (offshore) Dr Anil Malhotra, a promising and outstanding technocrat had shown me the poor response to our advertisement for recruitment of scientists for R&D work on a contractual basis of 3 or 4 years. The need for R&D institutes to be established was urgent. I had to overrule the laid down R&D regulations and created vacancies for regular appointment.

It is essential to have proper systems and procedures but one should not be a slave to them. When contingency demands in the interest of the organisation, the leadership has to overlook the norms and act.

13

Restructuring and Organisational Changes

The organisation at corporate headquarters was based on a mixture of functional and territorial basis, i.e. finance, personnel, material, exploration, onshore and offshore.

A major change was brought about in the organisational structure based on business group concept (Exploraton, Production, Technical Services and Drilling with support from Finance and HR departments). Six regions were controlled by regional directors who were made responsible for the performance of their particular region. Each member including finance and personnel was made accountable for the performance of one region each. **This brought about interdependence between the members, which resulted in smooth operations without conflicts.**

Organisation Development

As mentioned earlier, there was not even one member of the Commission who had risen from the first batch of executives. Members, including the chairman, were being inducted from other organisations. Executives were stagnating for years in the same positions. An organisation, whose core business was exploration, did not have a member exploration for 10 years! The morale obviously was very low. There was a brain drain from vital disciplines of management for greener pastures abroad and to other organisations.

The Twenty-Year Perspective Plan based on the hydrocarbon resource base within the country and opportunities abroad, brought to focus the need for organisational restructuring and development. In line with my commitment, when I retired in December 1989, every member of the commission had risen from within the organisation. The position of the vice-chairman in 'A' schedule, as of mine was created and filled. The individual was trained and groomed to take over from me in line with the succession plan.

All deputy general managers for promotion were interviewed and subjected to psychometric test. **A mutual assessment test between themselves were conducted to rate each other as friends, professionals and leaders. The subsequent growth of these executives did prove the merit of the system followed to rate them.**

In specialised areas, induction at various levels was done to bring about a balanced availability of expertise in various levels. Induction at the first level of executives was undertaken in line with wastage and future growth needs. A very well-structured training and development programme was laid down for the new entrants, particularly young executives. Apart from enhancing managerial and functional skills, a lot more stress was placed on personality and leadership development. They were encouraged to play team sports, undertake mountaineering and other social activities in the villages to develop stamina, team working through individual strengths, emotional intelligence and human relations through collaborative approach.

Consolidating Administration and Encouraging Welfare Measures

It did not take very long, as a result of measures enumerated literally on war footing, to ensure the total integration of rank and file as one family to ensure growth of the organisation with stability and continuous improvement in productivity. It was essential to further enhance the name of the organisation by being a role model, by meeting certain social obligations to the people of the organisation, retired employees, families of deceased employees and society at large. In this respect my wife Shobhana Wahi did a sterling work and played a selfless role to maintain a very peaceful and cordial environment.

In its areas of operations, including Dehradun, ONGC was an island of prosperity with poverty all around and had the obligation to provide some comfort all around. My wife set up Mahila Samities in Dehradun and in the regions for social and welfare work. Villages were adopted by Mahila Samiti's and a Gobar Gas plant was established in one of the villages around Dehradun. The Samiti also helped certain social bodies like Mother Teresa Homes.

Vocational centres were established to train the disabled and wards of employees.

My wife actively pursued and established a polytechnic at the National level for ladies at Dehradun (now known as Negi Polytechnic). Under the guidance of my wife and other ladies, it achieved a high standard of education in subjects like designing, computerisation, fine arts, fabric painting, fashion designing, etc., besides teaching the Montessori system of education. It proved to be a boon for girls who had to go to metropolitan towns to acquire training at great cost and discomfort. She also established a primary school called 'Sishu Vihar' for the lower income groups. Appendix K details out the work done for 'Sanghe Shakti' (Appendix-K). To bring about social interaction, a club for executives and a community centre for the staff were also built at Dehradun. As a social obligation to the area of our operation, the civil administration at headquarters was provided financial support for street lighting, sports and medical services (also refer appendices O and P).

To enhance the morale of the employees it is important that their physical, mental and psychological needs are looked after. With this end in view, a state of Art Hospital, community centres, an officers' club, etc. were established.

An area of 110 acres of land was acquired to build a very modern sports complex (unfortunately after my retirement the land was not used for the purpose it was acquired). For retired personnel, an extremely beneficial and novel medical scheme was introduced to take care of the medical needs of their spouses and themselves, at a scale similar to the one obtained during their service period. As a monetary benefit to the retiring personnel, a 'Good Health' scheme was introduced, through which they could encash a part of the balance of their half pay leave.

Consolidating the Work Area

The corporate headquarter was spread over a few buildings and numerous barracks, each having its own culture and work environment. It was decided to demolish all the barracks and build two multi-storey buildings to reduce movement of people and files and bring about a common work culture. One multi-storey building was completed to lodge the office of the members of the Commission and other senior staff.

Integrating On and Offshore Work Cultures

Senior executives were moved between the two organisational groups, i.e. onshore/offshore. There was resistance to these moves due to strong links with influential vested interest. In one particular case, I started getting pressure through the government. I had to tell them that they had a choice, either to move me out of ONGC or allow the senior executive to move from offshore to onshore. The officer was made to move. Brilliant officers like Mr Atul Chandra were shifted from Bombay Offshore to Assam. Some outstanding executives like Mr A. S. Soni were shifted from onshore to Bombay offshore. These officers brought about new ideas and dramatic changes in the work culture and performance in both the areas.

To quote one instance: Assam was facing severe land acquisition problems for drilling development wells. When an outstanding driller Mr S. M. Malhotra was transferred from Bombay to Assam, he took to inclined drilling from the same location, as fish to water. This resulted not only in tremendous increase in 'cycle speed', but also in great reduction of costs. A number of job rotations were ensured to prepare the executives to share higher responsibilities. Mr L.L. Bhandari another brilliant geoscientist was positioned from offshore, as director institute of petroleum exploration to give practical orientation to R&D work. Most of these officers rose to the board level positions. The seniormost geoscientist was positioned as member (exploration) within two months of my taking over as chairman. This ensured synergy between various disciplines within the exploration group.

Personal Example Set by Seniors

A new culture of mutual trust, understanding and respect evolved out of the integrated organisational structure. The new organisation structure was a threat to some of the members, particularly member offshore and member material, as their responsibilities were being restructured. This resulted in delays in getting clearance from the government. Member offshore historically was located at Bombay, whereas the headquarters of other members were at Dehradun. Member material operated mostly from Delhi.

On approval of the reorganisation proposal by the government, the headquarters of member offshore was shifted to Dehradun. Within 24 hours of the issue of this order, I was called by the minister. He asked me, "Colonel, how have you shifted member-offshore to Dehradun?" I replied, "Sir, by issue of an order." Next question was, "Why did you not get my approval?" "I did not want to bother you as it was within my powers," I answered. When he asked me why I couldn't retain him in Bombay, I explained to him that members have to operate from the same headquarters like the union or state cabinets. The matter rested there.

> A new culture of mutual trust, understanding and respect evolved out of the integrated organisational structure.

To mitigate the hardships to the families, as a result of transfers, special dispensations even outside the laid down norms were ensured. The morale and motivation of employees had to be ensured to enable them to perform beyond their normal potential. The new organisational structure helped to integrate offshore and onshore cultures. There was free flow of technology and people to different functions and business groups. Each functional member accepted the responsibility to upgrade technology and continuous development of people, to achieve excellence in operations.

It was our intention to promote ONGC as an integrated oil company to cover upstream and downstream. The organisation based on business group concept would lend itself to easy break-up as subsidiary companies for effective control as separate businesses.

Upgrading Performance Standards

The paramount importance to have a sensible and pragmatic approach towards training and development was realised. To meet these needs for improving performance standards, training programmes were upgraded and schemes to improve efficiency and productivity were worked out. Training for enhancement of knowledge, specialisation and development of management skills became a serious concern.

At the time when I joined ONGC, there were hardly five executives who had attended any major management training programme. Among those who had gone through management programme, most of them were considered as non-entities continuously involved in training programmes or temporary duty

DATA collection—(daily and travelling allowance). Arrangements were made with the Administrative Staff College, Hyderabad (ASCI) for managerial and leadership training of our executives in batches. A very intensive development campaign was launched. Initial reaction of some of the executives was that they were doing a good job and did not need any training. After attending the programme at ASCI, the feedback was extremely positive. They felt that they should have undergone such training, 10–15 years earlier.

Management Development

A management development institute at Dehradun (presently known as Academy) and staff institutes for training of supervisors were established in the Regions. A development culture was created, good libraries were established in the Regions. Thirst for acquisition of information and knowledge was created, through intensive personal interaction in periodic meetings and DO letters to the executives. All those who aspired to be leaders had to continuously update the knowledge to remain effective. In the annual appraisal forms, the executives were required to list the publications they had read. **It was well accepted that anyone who ceases to improve ceases to be good.** In organising these training programmes, several new innovative measures were adopted by getting institutional assistance from several training organisations of the country as well as from NRIs. Training and development of its executives and other employees became a top priority commitment for ONGC.

Delegating Powers (Empowerment)

Powers were revised to delegate them down the line to all those who had operational responsibilities. I had distanced myself from all purchase committees and even from the steering committee. The steering committee had two joint secretary level officers of the government as its members, along with the members of ONGC. The basic concept was centralised policy-making and decentralised administration. Periodic reviews were undertaken and high targets of performances were laid down. Futuristic planning was part of corporate responsibility. During situations of crisis, the chairman had to lead from the front.

Outsourcing—Cooperative Movement

The ONGC was operating from the lowest (sweeping) to the highest technologies, resulting in spending maximum time on the low technology areas and not on the core areas of the main business of oil exploration and production. It was felt that even though we have achieved manifold growth in the last couple of years, the potential for further growth of the Commission is very high. This growth could be achieved with stability if we concentrated only on high technology areas and farmed out low technology to trade, or by forming cooperatives with the help of retired employees and those who wished to take premature retirement.

> This growth could be achieved with stability if we concentrated only on high technology areas...

In keeping with this policy, drastic actions were taken to farm out low technology areas by forming cooperatives. The employees were motivated to take premature retirement with the incentive of getting on loan the equipment from ONGC. Shot hole drilling, transport, housekeeping, catering, etc. were outsourced. The cooperatives were provided technical and financial support. By adopting these initiatives the productivity and quality of work improved in every aspect of the business. In seismic data acquisition, improvement was almost four times in some areas due to outsourcing of shot hole drilling.

With this in view, the Commission also decided to use the helicopters of Pawan Hans Limited, wherein ONGC had equity participation to the extent of 49 per cent. Similarly, a separate company was formed by ONGC and SCI on joint venture basis, which would own and operate OSVs. A few OSVs then owned by ONGC were sold to private companies thus farming out totally the marine activity to SCI and the private sector.

Indigenous parties were encouraged to undertake more and more share of activities of drilling operations both onshore and offshore. Services like mud logging, seismic data acquisition, electro logging, etc. were farmed out. This reduced the capital acquisition and growth of manpower on low technology and support services within the Commission. The formation of cooperatives by unemployed youths particularly in the eastern sector was also encouraged to reduce pressure on the Commission for employment. The teeth to tail ratio in manpower was controlled.

Bringing in Transparency and Workers' Participation in Management—a Novel Experiment

A unique experiment in the history of the corporate sector in the country has been the formation of Advisory Committees to the Chairman, on Exploration, (Appendix S1-S2) Human Resources Development and Management. The members of these committees were eminent professionals including senior retired executives of ONGC, Director of Management Institutes, Media personalities.

The presidents of all unions were members of the Advisory Committee on Management. The interaction with them was of great assistance, particularly in the management of change and launching of new initiatives. The members of forum were eminent people in their own rights. They could ask any question and give feedback from the business and political environment. This also enabled ONGC to achieve transparency in its working, particularly in matters of high value purchases and projects. The members of forums also acted on their own as spokespersons to manage the business and political environment. This enabled the smooth management of radical changes brought about to achieve organisational excellence.

This further enriched ONGC's management philosophy and ensured full involvement of the labour and intelligentsia in the management's decision-making. It had an immense impact on the relation with employees, who in effect became a part of the management, and the dangerous spirit of confrontation with the management generally found in enterprises dissipated away. A collaborative culture got created. It was no more a game of football between the two to win. On the contrary, the workers became a part of an enlarged team, working to win the goal together. ONGC was the only organisation in the public sector which did not go on a five-day week which even the government had introduced. There was not the slightest murmur on this account till 15 December 1989, when I had demitted office.

Farming out of low technology areas, outsourcing a number of activities which were earlier being handled in-house, was seen as a threat by the workers. Apart from meeting the unions to explain that the existing people in the organisation will not be laid off as the organisation was on a growth path, the novel method of ensuring workers' participation removed any doubt of retrenchment from the minds of the workers and guaranteed their wholehearted commitment for the organisation's progress.

Improving Communication and Logistics

Dehradun as a headquarter had no air link, roads were not fit for fast movements, and rail movement was slow. There was not even a hotline between Dehradun and Delhi, let alone the fact that the areas of operations were widely spread all over the country, in the north, south, east and west, both offshore and onshore. The radio link with Assam was temperamental and obsolete. Where a dynamic organisation needed live data and information for live support, the information at the headquarters was received only after the events had occurred and at best was merely used for a postmortem. There was no sense of urgency and a sleepy and status quo culture prevailed. The senior management supervision and support in the Regions, which were spread over wide and long distances, was also inadequate as movement was by road that was tiring and only meant the loss of management time.

To improve communication, Indian Airlines was persuaded to start air service to Dehradun. Helicopters were provided to the Regions. In spite of bureaucratic hurdles, a six-seater executive Dornier plane was bought for senior executive movements, within the country. A state of art communication system was established which enabled Dehradun to be the centre of activities and the corporate headquarter.

As indicated earlier member (offshore) who was historically located at Bombay was shifted to Dehradun, which gave all the importance and support that the headquarter deserved. Appendix P gives a glimpse of the assistance extended to Dehradun town.

Enhancing Work Process through Computerisation

Prior to my joining ONGC, there was only one computer centre at Dehradun for processing the seismic data. With the explosion of seismic data collection by 2D and 3D seismic work from all the Regions, the capacity of this computer was woefully inadequate. Commercial processing was done abroad by companies like DIGICON, Western Geophysical or CGG which were not always target oriented. This also had the risk of data leakage. Building up viable geological models required target-oriented data acquisition and processing and interaction between the processing centre and the field operator. Besides enlarging vastly the capacity of the computer centre at Dehradun, regional computer centres were also established. Interactive work stations were introduced which revolutionised the quality of the interpretation and of the 3D data. The communication system was modernised with the latest available technologies.

Computers were also exclusively acquired and installed in the Institute of Reservoir Studies for Reservoir Modelling. Mini computers were installed for processing of electro-logs, geochemical and geological studies. Commensurate with the huge collection of data, the data processing set-up in all areas was upgraded. Executives were also provided computers on their desks to help them store and use data expeditiously. A major thrust on information technology was given.

A modern computer center building was designed and constructed, by making ONGC Civil Engineers to visit some computer centers abroad.

*****

Organisational changes in all spheres triggered off revolutionary growth. A new culture had emerged. A motivated workforce and a committed management had evolved. A large number of discoveries, not merely conventional but also stratigraphically complicated ones, were discovered. The performance charts showed a remarkable rise. (Appendices G & H). The ONGC had achieved international recognition. India as a result found its rightful position on the world oil map.

Dehradun, the head quarter of ONGC, became a favourite place to visit by ministers and delegations from developed and developing countries. ONGC was favoured by the visits of our two presidents, prime minister, chief ministers of states and many other dignitaries (Appendices E, F, Q).

14

Bombay High Oil Field – Canards

Bombay High Oil Field is the largest oil field in the country. Bombay High structure was delineated with the help of a Russian vessel – *Academic Arkhangelisky*. The first well drilled on the structure by the then recently acquired jack-up rig *Sagar Samrat* discovered the field on 19 February 1974. The Bombay High oilfield structure is quite large by any standards and is located at about 160 km off the shores of Bombay. The depth of water at the drilling locations ranges from 75 to 90m.

Soon after the discovery of the oil field in 1974, the presiding incumbent of the organisation, B.S. Negi who had officiated as chairman for around four years, retired from service and was replaced by a permanent chairman. The management under the chairmanship of Mr B.S. Negi was conservative but highly committed to geo-scientific professional expertise, which had resulted in the discovery of Bombay High Field.

I had the pleasure of interacting with him a few months after I had assumed the charge of ONGC. The results of the SWOT analysis of ONGC and the outline of the long-term plan prepared by the task force headed by Mr R.N. Misra was explained to him. Negi was appreciative of the efforts and felt that the changes we were planning to bring about were radical, nevertheless he wished us luck. He however expressed his reservations on the strategy followed immediately after his retirement for the development of Bombay High Field when, for reasons not recorded, production from the Field had acquired a new urgency without making proper reservoir assessment. He felt that a lot more delineation wells should have been drilled to fully appreciate the extent and potential of the field before putting it on such a scale of commercial production. This observation was well accepted, which indeed was also the viewpoint of many of our geoscientists. This approach was followed, in tandem, during my chairmanship that did give rich dividends.

Dependence on Foreign Consultants

As mentioned earlier, at that time there had been a very strong culture of dependence on foreign contractors for the development of Bombay High Field and associated offshore platforms and pipelines. As a consequence of these activities the development of indigenous capabilities within the public and private sectors got suppressed. The reservoir experts were of the considered view that the development of the Bombay High Field was undertaken without thorough delineation of the field and without laying out proper detailed long-term development plans for pressure maintenance.

It may be appropriate to mention here a few words about the CFP contract. A mutual understanding between the French and Indian governments for cooperation led ONGC and CFP to sign an agreement for the development of Bombay High. When I took over the chairmanship of ONGC, I observed that the commitments made by CFP at the time when they had made a bid for the development of Bombay High were also not fully realised. CFP contract was being operated from Bombay without the desired level of involvement of the research institutes and the transfer of technology to ONGC. I therefore decided to post a senior officer of the ONGC (on a rotational basis) at CFP's headquarters in Paris for the duration of the contract to ensure the desired level of involvement of the ONGC geoscientists and engineers and to ensure transfer of know-how.

The scheme for in-fill drilling to improve the recovery factor by about 10 to 12 per cent was finalised through a brainstorming session with our geoscientists, which was very reluctantly accepted by CFP. In fact, we had to send a team of our geoscientists headed by our brilliant geologist Dr S. Ramanathan to CFP to make them understand and accept the scheme. My contacts and friendship with the top level managers of oil companies was also of great help in getting cooperation. Amongst them Mr Deny, Vice Chairman of TOTAL an outstanding professional was a great help to get the best service from CFP till the contract lasted.

Production from Bombay High giant field was under constant review to ensure best known production practices. A very capable reservoir scientist from the Institute of Reservoir Studies (IRS) was stationed at Mumbai. He was made accountable, for the established reservoir management practices and his decisions to the Director, Institute of Reservoir Studies. In an attempt to ensure that the reservoir management of the offshore field does not suffer on account of any lack of knowledge, I suggested to Mr JMH Van Engleshoven – MD SHELL, during one of my meetings, for technical collaboration on R&D. His reply was "Neither for love nor for money but since you are friend, I shall allow your Scientists/ Engineers to visit our R&D installations. They can discuss the problems, and a lot of technical know-how can get transferred. Thereafter, I shall send a group of our scientists/engineers to India on problems identified for discussions". A team from SHELL did visit us, after our team had returned from the visit to SHELL Laboratories. SHELL team was requested to review the management of Bombay High Field apart from other issues in Assam and Gujarat. Extract from the conclusion of the report by SHELL team consisting of M/s D. Antheunis, C. Van Barren C.De Nsaard and headed by Mr K.G. K. Menon, one of the best reservoir engineer in the world, is as follows:

> The manner in which the giant Bombay High field has been rapidly and efficiently developed by ONGC appeared impressive. With the pressure maintenance scheme (water injection) that is already in operation and the sound reservoir management measures that are being taken by ONGC, it appears that the present off take level of under 400,000 stb/d is somewhat conservative as judged against the estimated recoverable reserves (6% p.a. depletion only).

The status of production in Bombay High was discussed in the meeting of CACES (Chairman's Council on Exploratory Strategy) held on 16–17 November 1988. The presentation was made by Mr P.K. Kulkarni, general manager (reservoir) in Bombay Offshore. An extract from the minutes of the meeting is reproduced below.

> In view of regular monitoring, rescheduling of production, start of water injection in up dip row wells, successful work over operations, introduction of innovative temporary gas lift measures and sound reservoir management practices have indicated that Bombay High field is behaving better as confirmed by reservoir simulation studies by IRS in March/April 1988.

This meeting was attended by ONGC members, former (retired) senior geoscientists and among others, late Lovraj Kumar, A.B. Das

Gupta and Dr Hari Narain. They had all appreciated the excellent reservoir management of Bombay High. On this occasion, I further detailed out the strategy followed for the effective management of the Bombay High Field.

Keeping the present oil production profile as implementable scenario, an optimistic production scenario wherein all such likely extensions and accrual from additional discoveries over Bombay High are taken into account shall be built up. It will help in identifying the probable production potential of the field.

A task force shall be constituted which will constantly monitor the reservoir performance, study input requirement, implementation and suggest mid course correction. This group shall also undertake future planning and decision analysis for a better Reservoir Management of Bombay High as our endeavour has to be to maximise the recovery from this 'Giant'.

In spite of all the scientific measures adopted in the field and the technical suggestions provided by the best people available in the world, a whispering campaign started on the over-exploitation of Bombay High, by vested interests. This, besides other motives was deliberately initiated to attain the consultancy assignment. Greed has no limits. Success gives birth to jealousies and detractions. It was possibly in this vein that a joint secretary in the ministry of petroleum once made the mistake of suggesting to me about the formation of a committee to examine the performance of Bombay High. He was consoled by bringing to his knowledge that all the activities at Bombay High were constantly reviewed by associates as well as external experts including retired eminent ONGC scientists. I objected to the formation of any such committee by the government observing that such a committee will only demoralise our geoscientists, which I told him is not acceptable to me. It had taken many years to build an excellent team and I would not allow anyone to destroy it, as long as I remained the Chairman.

This canard of over-exploitation took the main stage, a few months after I had retired (15 December 1989). I had handed over the charge to Mr P.K. Chandra who had been elevated to the position of vice-chairman and had been groomed to shoulder the responsibilities as chairman. Alas! the detractors had other plans to destroy ONGC and managed to bring a non-technical person to take over the chairmanship of ONGC. In this context I am tempted here to narrate the following anecdote.

> On the retirement of a CEO, the next incumbent who had reached the level of his incompetence, asked for his advice, in case he faced any problems. He was given three envelopes marked one, two and three, to be opened during subsequent problems. In his inability to maintain the performance, he opened the first envelope and found the advice 'Blame the predecessor'. After a lapse of another few months, he opened the second envelope which contained the advice: 'I am reorganising and inducting vigilance officers to help'. On reaching the disastrous situation, he opened the third envelope and the advice was—'Write your three envelopes'.

During the tenure of the new chairman there were all-round slippages in production and work output tempting him to take the easiest recourse to the problem by blaming the practices during his predecessor's time.

The following extracts throw light on the state of affairs then existing.

Too Many Idle Wells Depress ONGC Output—Worrisome Trends on Oil Front

'Over a third of the existing and newly drilled wells of Oil and Natural Gas Commission are not producing oil and the high level of idle capacity has been caused by lack of management inputs and non-availability of equipment to stimulate the wells,' says T.S. Subramanian.

The Oil and Natural Gas Commission is at the center of a controversy following a sharp drop in the output of petroleum crude and a much lower target for the current year. Both have serious implications for the economy as the balance of payments position is substantially linked with indigenous availability of oil. Lower domestic output has compelled increased import of oil and products, worsening the foreign exchange crunch.

The ONGC slide in performance began when its production fell to 30.35 million tonnes in 1990-91 from 32 million tonnes in the previous year. There was also underutilisation of natural gas. All this led to a reduction in profits by about Rs 560 crores. ...

It has been a repeat performance for 1991-92. The Working Group of the Planning Commission had set the target for the current year at 34.1

million tonnes, which was brought down to 32 million tonnes. But on April 6, 1991, the ONGC informed the Centre that it was further reducing the target to 28.17 million tonnes (Bombay High—18.20 million tonnes, Gujarat 6.5 million tonnes, Assam 3.15 million tonnes and southern region 0.31 million tonnes). ...

The present management also claims that the declining production of crude is mainly because of the damage to the Bombay High reservoirs due to 'over-exploitation' by the previous management. It readily cites the A.B. Das Gupta committee report to buttress its claim that the previous management (under the chairmanship of Col. S.P. Wahi) had overproduced from the Bombay High and as a result impaired it.

The Government of India, through a memorandum dated April 26, 1990, constituted the committee to review the development of Bombay High reservoirs. ... The committee also went into the question of whether there were any deviations from the production programme in the Bombay High reservoirs, resulting in impairment to them. (This committee was constituted after the retirement of the author).

'Compromises' in Panel Report

Oil experts, however, assert that facts speak otherwise and that oil production has come down not only in Bombay High but also in Gujarat, Assam and the southern region (Cauvery Basin). They say the Das Gupta committee report is full of 'compromises' and that its technical findings run contrary to the conclusion arrived at. Giving an example of how it has hedged its bets, they point to the statement in the report which says, 'The deviations in production both in Bombay High North and South were justified according to the accepted industry practices to meet production commitments. However, with regard to the point of view of maintenance of reservoir health, the deviations were not justified.'

The basic reason for the falling production is that out of ONGC's 3,500 wells, about 1,180 are idle that is, they are not producing oil. ...

'Thirty-seven per cent of the wells were idle in the ONGC in 1980-81. It was brought down to eight per cent in March 1990. But by March 1991, it had gone up to 38 per cent. This explains why the production is short. That is where the crux of the problem lies,' assert the Commission sources.

It is pointed out that the substantial idle capacity has been caused by the lack of management inputs and non-availability of equipment to stimulate the wells. The ONGC employees complain that the top brass do not visit the drilling sites or group gathering stations to find out the problems that exist at the grass-roots level. 'We do not have casing pipes, gas lift equipment, etc. The Chairman has not bothered to go into the grass-roots problems. He thinks it is the problem of various regions alone,' they allege.

Source: T. Subramanian, 'Too many Idle Wells Depress ONGC Output', *The Hindu*, 10 October 1991.

Bombay High Has Not Been Flogged

In its report which came as a relief to the country's jittery oil managers, the consultant has packed compelling arguments to prove that Bombay High has not been flogged or damaged; it has not lost oil either.

Gaffney Cline has debunked many theories on what ails the country's premier oil field including prescriptions by a few oil experts.

First, it says it is incorrect to compare Bombay High with North Sea or any other large field. Bombay High, with its unique characteristics, is far more complex than North Sea Oil Fields.

Reducing the rate of production or closing down some of the wells with high gas oil ratio, it says, is not the solution. It feels the present recovery level of 26 per cent can be raised to 35-40 per cent with proper investment.

The 1,200-sq km Bombay High is not comparable with North Sea whose fields such as Brent, Forties and Ekofisk have areas ranging from 50 to 90 sq km. In North Sea, one well has a producing zone which is 5 to 16 times bigger than Bombay High, while Oil platform at Bombay High has 26 million barrels of recoverable reserves. Its counterpart in North Sea has 440-1,400 million barrels.

Source: By R. Sasankan, *Business Telegraph*, Calcutta, 16 December 1998.

The story of mismanagement will be more clear from the following extracts taken from I.A. Farooqi's *The Story of ONGC*..

Immediately after Wahi's exit, Commission started downgrading its work targets and the Government of India was approached within two months of his departure to approve slashing of annual production target. The concerned department ascribed the expected short fall in the production to the slippage in the completion of engineering projects. The weak and humble leadership at the top in ONGC, instead of driving the people hard enough to accomplish the target resigned submissively to the proposal.

During this period, the overall production of oil fell and even the reduced yearly targets were generally not achieved. Against a production target of 33 million tonnes of oil in 1990-91 only 30.35 million tonnes were produced. In 1991-92 the targets were further scaled down to 29 million tonnes but production of only 27.81 million tonnes could be achieved. In the year 1992-93 only 24.43 million tonnes were produced against the target of 25.54 million tonnes. The reduction in the production targets was affected mainly in the Bombay Offshore area, on the plea of repairs of some of the wells. A well orchestrated fault-finding process of earlier production rates was started and the charges of flogging the offshore wells (production at a rate higher than technically permissible) in earlier years was freely floated, around which generated quite heated discussions in selected circles. However, contrary to the ongoing criticism, the World Bank technical team that visited ONGC during that time was of the view that the Bombay High Field was being under produced. The Shell Group team that had examined the production data during Wahi's time had also expressed a similar view. A major cause for reduction in the offshore production of oil that fell to the low level of 18.9 MMT in 1991-92 was due to slippage in many of the engineering contracts and delay in repair of sick wells.

It is a pity that some of our own so-called eminent people show such lack of intellectual integrity for personal gain and politicise technical issues which had a tremendous demoralising effect on ONGC's outstanding geoscientists for whom I have the highest regard.

Effective Management

15

Leadership and Communication in Crisis Management
(Sagar Vikas Blow-Out)

Crisis situations are a part of the game of oil exploration and production, mostly connected with blow-outs. Blow-out is the most dreaded operational hazard in oil exploration. It generally occurs when the formation pressure greatly exceeds the pressure exercised by the drilling gear and the chemical mud that is pumped in as the hole is drilled. This leads to gushing out of the formation fluid when the force cannot be contained. The real test of the team work is exhibited during such crises. It brings to light some useful lessons on human psychology, the importance of communication and the role of prompt decision-making. One such incident occurred on 30 July 1982 at 9 p.m.

On that fateful night uncontrolled flow of reservoir fluids—gas, muck, traces of oil, stones came gushing out of well no.5 at SJ-Platform, which was being drilled at a depth of 1,660m by the jack-up rig Sagar Vikas. The site was located 78 nautical miles off Juhu coast in the Arabian Sea on Bombay High structure. Noticing the situation going out of control, the crew of 74 was safely evacuated and ferried to the Sagar Pragati, which was about 5 km away from Sagar Vikas.

The credit for this safe evacuation goes to the excellent systems which had been laid down by member offshore.

Reactions to the Crisis

- Media became restless for the story beyond the story 'given out'.

- Government, worried at the unfortunate accident of this magnitude, sought full details from ONGC.
- The ONGC family was stunned and gloom cast all over the organisation.
- Parliament (which was in session) maintained a dignified calm and waited anxiously for the petroleum minister to make the statement.
- Ecologists got worried about the pollution of the sea.
- Public in general wanted to know how it happened.

Participative Communication

For keeping up the morale of the staff it was essential that the news reached ONGC employees at various work-centres all over the country through the media in a correct perspective. Distortion of information could create demoralisation in the rank and file, which would be inimical to the interests of the organisation at that crucial stage.

As an extension of participative management, it was essential to involve the media, particularly the mass media, in an effort to share information not only within the largely scattered population of ONGC, but also with the public in general. It had to be an exercise in 'participative communication'. It was an innovative necessity borne out of the anxiety to hold together, in the face of an unprecedented situation, both internally as well as externally.

I held a press conference and gave a commitment to meet them every day and share information and documents asked by them. Even the experts or people of the ONGC they wished to meet were made available. They in turn promised that no unconfirmed news would be published. It was an open and transparent agreement which stood the test of time.

Communication Strategy

A strategy was thus evolved to achieve broad communication objectives.
- No room to be allowed for rumours.
- Top ONGC management to become the spokesperson during the crisis.

- Regular media briefings to be held in the evenings at the offshore headquarters in Bombay.
- Press notes to be issued regularly and relayed to all the major centres at work.
- Media to be given full facts and figures as well as any assistance in filing their reports.
- Information levels of the eagerly curious media men to be upgraded particularly regarding technicalities involved in the blow-out.
- Media to be taken to site for a 'fly over' as soon as helicopters became available.
- Public relations office to remain in constant touch with the Radio-Room and collect eyewitness accounts of the happenings at the site from those reaching base.
- Provide evidence, as far as possible, to substantiate statements made.
- Monitoring of published media reports, particularly those not based on facts. Immediate clarifications to be given at the evening's media briefing.
- Personal visits by top management to other installations to directly communicate facts.
- Blow-out and subsequent activities to be timed for future reference, training and education.

Implementation of Communication Strategies

Media was briefed every day by the Chairman. All the requirements of the media regarding visit to the site, meeting with Red Adair team or examination of documents, were met. The media as promised gave correct and factual reporting.

Report to the Nation

A report to the nation on the crisis was conceived in the form of a telecast at the peak hour—10 o'clock news, which had the blow-out as the main headline. Instead of the newsreader narrating the account, the Chief executive of ONGC, in an appearance which lasted about 10 minutes, enumerated the steps being taken to control the blazing well. This had a reassuring effect on the general public. In Parliament, the minister for petroleum made a statement on the blow-out, giving full facts of the accident and steps being taken to control the situation. The minister's statement was received with sympathy and understanding. The nation was now behind ONGC.

*****

I had reached Bombay by the first available flight. On landing at the Sagar Pragati helideck, I was met by Dr A.K. Malhotra, Mr Woodward and Mr Cheema. I shook hands and said, "Let us have a cup of coffee." They looked at each other. I could see from their body language, a feeling of disbelief. They obviously thought that the chairman was clueless about the danger and the crisis at hand. They were expecting obviously a stiff reaction (wrong perception of army officers), whereas I was worried about the tension under which they were operating and wanted to diffuse the situation. When I reached the rest room, I heard a number of voices from the evacuated crew trying to explain as to how the blow-out had happened, and the person responsible.

I asked, "Who was on the press brake?" There was pin-drop silence as they felt I had caught the right man. Someone murmured, "Manmohan Singh." I said, "Where is he?" I could immediately notice a sigh of relief as the focus had shifted to Mr Manmohan Singh, who was resting in the bunk house. They said, "Should we call him?' I said, "No, I will go there." There was fear on their faces as they thought the worst for Manmohan Singh. When I reached the bunk house where the poor fellow was lying, he jumped up, as if he had seen the ghost. I told him, "Relax, I have come only to see you, as you had faced the brunt of the blow-out. Nothing will happen to you or anyone." In fact, no one was even asked to explain. A fact-finding committee was ordered to take corrective actions if required for the future.

The following issues needing actions were identified:
- A generator on Sagar Vikas was still running and could ignite the gas flowing from the blow-out well.
- To control the blow-out, 'killing' (making it inert) and caping of the well was necessary.
- Morale and motivation of the employees were to be kept high all over the country, so that operations in other areas were not effected.
- Likely pollution was to be safeguarded.

- Parliament was in session and had to be kept informed through the ministry to get support.
- Media to be managed by sharing information with the public.

> ❗ It was a dare devil action fraught with danger of fire at any time. ❗

Switching off the Generator

To switch off the generator, helicopter pilots, who were Canadians, had to be motivated to land on the leg of the rig as the helideck was covered by muck and gas. It was a dare devil action fraught with danger of fire at any time. In such crisis situations, risky decisions are required to be taken. Instructions were issued to keep a record in the control room of the authority taking the decision for review after the crisis. This had the desired impact in the government and the following message was received from Petroleum Minister, Shri P. Shiv Sankar.

> I made statements today in both houses of parliament regarding the blow-out that had taken place on the SJ platform and Jackup Rig 'Sagar Vikas'. Both the houses have expressed their best wishes and full support to the ONGC in their efforts to control the Blow-out successfully and early. My personal wishes and support along with the views of the two houses may please be conveyed to all concerned in the ONGC. Wish you all success.

A news item is reproduced below which details the progress of events after the blow-out had struck.

ONGC's Amazing Feat in Face of Disaster at Sea

> At 9.15 p.m. on Friday, July 30, disaster struck. Well No. SJ-5, being drilled by the Oil and Natural Gas Commission's jack-up rig, Sagar Vikas, in Bombay High, some 100 miles from the shore, suffered a blowout. As oil and gas gushed forth uncontrollably from the stricken well, the danger of a horrific fire loomed large. But, within half an hour, all the 74 men aboard the rig were rescued and hauled away to safety. It was an amazing feat. Monsoon winds at the speed of 30 knots continued to lash the waters and raise 15-ft-high waves in the sea around.
>
> At 9.25 p.m., the telephone buzzed in the Altmount Road, apartment of A.K Malhotra, ONGC's Bombay-based Member (Offshore). Within minutes, he was in the control room at Maker Towers, Cuff Parade, directing operations. Thus within 4 hours, a supply vessel, Gulf fleet-46, was at the site, spewing a jet to form a water-screen over the threatened

rig. In less than 24 hours, it was joined by another multipurpose support vessel *Pacific Constructor*. This and the assembly of related equipment at such speed, again was a spectacular achievement. In the North Sea and the Gulf of Mexico, where the stakes are much higher, multinational oil giants have often taken anything up to a week to mobilize resources on this scale to fight a blowout.

Preliminary trouble-shooting over, Mr. Malhotra flew by chopper to Sagar Pragati, ONGC's nearest platform to Sagar Vikas, just 5 kms away. Within an hour, Mr. Cheema, General Manager joined him. All night, they assessed and battled the hazards. Around 6 am on Saturday morning, Col S.P Wahi, the Chairman was informed in Delhi. By 10.30 am, he too was at the site.

The main worry of the trio was that a spark—any stray spark—could set ablaze Sagar Vikas and the jet of Oil and Gas issuing forth from the 'runaway' well. So far, the water-screen had managed to put out every such spark. But clusters of cables hung from the bracings were continuing to rattle dangerously in high-speed winds. The 'wild' well might also be spewing stones that could collide against the metal casings or anything else to cause fire. So far ONGC had been lucky. But how long?

As it happened, only till Monday morning, when the dreaded inferno did occur and all but the stumps of Sagar Vikas slowly went down blazing into the sea.

But on Sunday, the generator in the deserted Sagar Vikas was purring as if nothing had happened and the lights aboard were ablaze. They had been deliberately left 'on' to ensure the safety of the disembarking crew on Friday night. The rig's helipad enveloped by misty gas was virtually invisible. So ONGC decided to take a leaf off the pages of science fiction and enact their own version of a spy movie.

In a daring operation, a helicopter landed on the left leg of the drilling rig, a man aboard got out and clambered some 25 ft down the derrick-like structure, switched off the generator, climbed back up and into the chopper, and was off. Col. Wahi, true to his army tradition, was the first to volunteer for the task but his men would not let him.

All in all, the disaster at sea has tested the nerve of ONGC's men and top leadership as never before and boosted morale across the board. Thanks to timely insurance, the commission may not suffer any direct financial loss. More to the point, it is confident of achieving, if not exceeding, this year's production target for oil and gas from operations offshore.

Source: *Illustrated Weekly*, 6 August 1982

President Dr Rajendra Prasad being presented by Capt Wahi capabilities of armour during a visit to 18 Cavalry in 1955.

18 Cavalry tank being recovered

With 18 Cavalry officers – Capt. Wahi standing 2nd from left.

Vijayanta tank – creation of Mr. Mantosh Sondhi

My mentors—two outstanding leaders: Mr. Mantosh Sondhi and Dr. V. Krishnamurty

Visit to North Sea (UK)

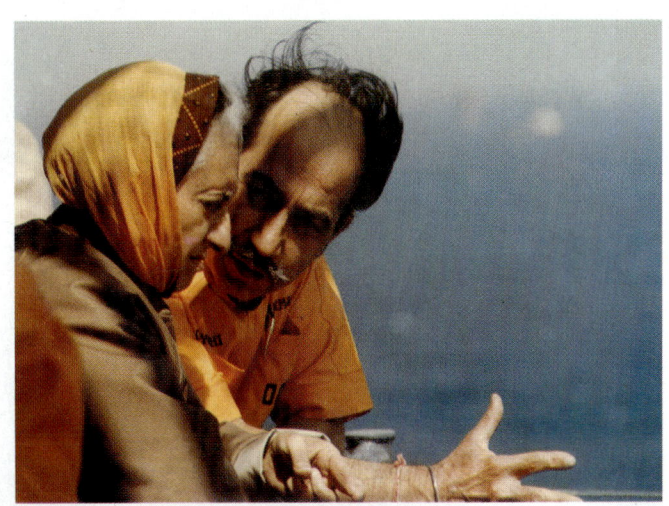

Mrs. Gandhi on a visit to Bombay Offshore with Col. Wahi

Dr. Manmohan Singh and other policy makers – a novel method of ensuring the commitment of the goverment.

Indigenous development

Mr. Deny, Executive Vice Chairman, TOTAL, giving an affectionate welcome salute in Indonesia.

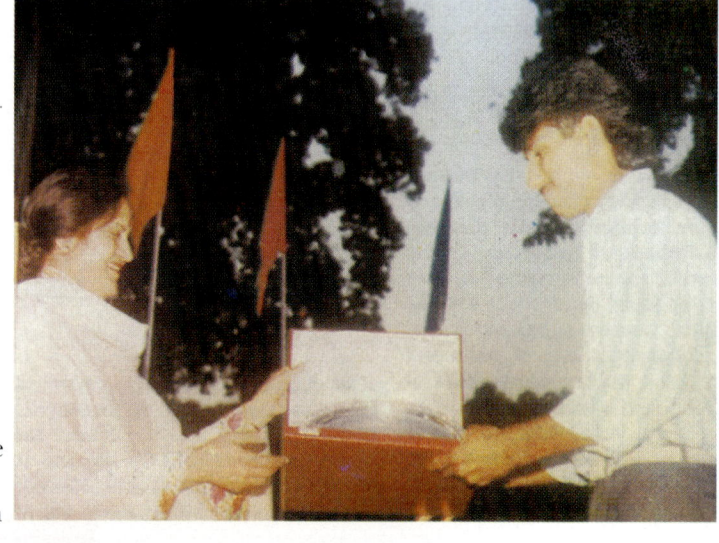

ONGC stresses on sports and cultural activities to promote integration and leadership development: Mrs. Wahi presents the amateur trophy to Chiranjeet Milka Singh

Col. Wahi and Mrs Shobhana Wahi with Mother Teresa, who blessed ONGC

Chairman and Member (Offshore), ONGC, debriefing the ONGC team (Chapter 15)

Reporting to the nation: initiating transparency and accountability (Chapter 15)

Dare devil landing of Helicopter on leg of an oil rig as helideck was covered by gas and muck. (Chapter 15)

Chairman Col. Wahi dressed as part of the crisis management team (Chapter 15)

Warm welcome in Nagaland to Chairman Col. S.P. Wahi (Chapter 16)

CHAIRMAN WITH HON'BLE CHIEF MINISTER OF TRIPURA

Col. Wahi with the Chief Minister of Tripura, relationship management (Chapter 16)

Mrs. Shobana Wahi, patron and founder of ONGC Mahila Samiti, adopted villagers around Dehradun for upliftment

Visit to Nagaland to meet the Cabinet (Chapter 16)

Visit of PM Rajiv Gandhi along with Col. Wahi to Assam

Col. Wahi on a visit to EME College, Trimulgery (AP)

Col. Wahi as president of the golf union making a 'tee off'.

Keeping the Morale High

The inter-regional cultural programme at Dehradun was not cancelled. I visited Dehradun for a few hours to personally brief the employees from the six regions about the incident and exhort them to carry on with their activities with even more vigour in this hour of crisis at Bombay offshore. The employees from the different regions had carried this message back straight from the chairman to keep the morale high.

Fighting the Fire

In spite of switching off the generator, the production platform and Sagar Vikas were engulfed in fire through a spark caused by friction between steel and stones which were being thrown up as part of the muck. Pacific Constructor an MSV (Multi Purpose Support Vessel) was mobilised to throw water to keep the structure and platform cool. This was essential as four completed oil wells were in danger of being damaged by the fire.

On reaching 'Pacific Constructor', during my visit to all the offshore installations, I asked for a cup of coffee and sat down. The captain of the ship, a young, tall and handsome Englishman came over. I could immediately see that he was furious. He said, "Mr Chairman, I am 33 years old and cannot live with the death of so many people on my head". I had anticipated this type of a reaction. I had felt the heat of fire while landing on the helideck of 'Pacific Constructor' in severe rough sea condition. The captain was obviously scared and was imagining an explosion. His deep sea divers were in the compression chambers and had all types of gases and explosives on board. So I looked up and told him, "Young man, I have 33 years of experience and I can see that you are having cold feet. I shall take over the command of the ship and relieve you". The chairman of the vessel was informed about the situation and his reply, through telex dated 5 August 1982, is reproduced below.

Colonel S.P. Wahi 5 August 1982

> I have tried to phone you without success. On behalf of all of us in Swire Pacific Offshore, I would like to express my sincere condolences to you and all in ONGC over the tragic Blow-out and fire on the rig 'Sagar Vikas'. I understand there was an outburst on the bridge of 'Pacific

Constructor' by Captain Stretch whilst under strain at the outset of the crisis, which is deeply regretted.

I believe, however, that fire-fighting has since proceeded smoothly and all on board are co-operating with ONGC to their utmost in trying to minimize the extent of the disaster. My deepest sympathies.

David Gledhill
Chairman, Swire Pacific Offshore

This brings out the importance of leading from the front. The captain had the intention of pulling away the MSV some distance away, but then throwing of water would have been only of ornamental value.

Relighting the Fire—Experts Overruled

Mr Henry, the Red Adair man was being briefed to meet the press. A message came from the 'Pacific Constructor' that fire had got extinguished.

I told Henry, "Please go to the control room and I will follow you after informing the press to wait till we come back." I had failed to identify the control room he had to visit. Henry was not there, when I had reached the control room. I was told that he had gone to the other control room and had given orders to relight the fire. I directed that no action will be taken to relight the fire till further orders. Henry and I went back to the control room. He talked to his deputy on the 'Pacific Constructor' who was ready to relight and confirmed his earlier recommendations. Dr AK Malhotra, member (offshore) too mentioned that his best judgement was to relight. However I was convinced that we should not relight. The sea was rough and the wind was very fast. There was hardly any oil being thrown into sea and unburnt gas was being blown away. There was no hazard with regard to pollution or to the safety to other installations. These were four completed wells on the platform. There was a danger that these wells may also get damaged with intense heat, resulting in irrecoverable oil. When I told Henry my decision, he literally looked down at me. Firstly he was taller and secondly he thought adversely of my decision. He said, "Well, you are the chairman." It was decided to visit the site by first light and in the meantime the MSV was to be maintained. That was the only night when I could not sleep properly as I had overruled the best known experts in the world, when my colleague was in favour of relighting.

Next morning we reached the site, Henry stood up in the helicopter, took his cap off and said, 'Chairman, you have saved at least a 100 million dollars for your country. Our judgement was wrong.' Everything got recorded by the control room.

Mr Talukdar member (exploration) in his fact-finding report had mentioned that it was the bold decision of the chairman not to relight, which saved a major disaster even though Dr A.K. Malhotra had sided with the consultants' view of relighting. Mr Woodward, the seniormost member, took control of the log book from the control room, and had discussed the subject with Dr Malhotra, who then wrote the letter dated 20 April 1983, which is reproduced below.

> Dear Col. Wahi,
>
> I have for reference the correspondence resting with Shri Woodward's D.O. letter No.5/11/82-Mon dated December 4, 1982.
>
> Since the matter of reignition was not germane to the Enquiry, this correspondence was really not necessary.
>
> As I recall, one could have equally gone with the advice of Red Adair and his Associate Experts to reignite but the decision you took not to do so did prove itself to be of far better merit as subsequent events showed.
>
> I shall, therefore, be grateful if you will kindly treat this matter as finally closed.
>
> With regards,
>
> Yours sincerely,
>
> Dr. A.K. Malhotra

As A.K. Malhotra was a brilliant technocrat and was being groomed to take over the reins from me, I had therefore overlooked the issue which had caused some embarrassment.

Tools Used and Gains Made

Bombay High blow-out was a threat turned into an opportunity by the deft handling of an otherwise perplexing situation. The main tools used in containing the likely blow-out of public opinion were an 'open house' policy. All baffles were removed. All plans, all developments taking place were shared with the media. The information given was based on truth and frankness.

The Sagar Vikas rig and associated losses were fully covered by insurance. The crisis, however, brought the employees together. The leadership demonstrated the ability to act with speed and courage in the interest of the organisation by leading from the front without worrying about personal discomfort and risks. It was a pleasure to watch with great admiration, everyone from member (board level) downward putting working overhauls to lend their hands to the task. The Public Relation Department headed by M.L. Kaul had done a commendable job in managing the communication.

An extract praising the ONGC's performance in bringing the blow-out under control is reproduced below.

> The Oil and Natural Gas Commission's performance in capping the blow-out SJ-5 well at Bombay High has been impressive. Within two hours of the blow-out on July 30, the entire top brass was at the site and the Red Adair Co. had been put on the alert in Houston, Texas. Within 24 hours, two fire fighting vessels were training jets on the well. Another 80 hours later, three Red Adair blow-out fighters had arrived. By all accounts, the ONGC's oilmen were able to cross bureaucratic hurdles in mobilizing the equipment needed to control the blow-out. Their plans, well conceived and realistic have won them international praise. The ONGC was also plainly lucky on two counts. The fire kindled by friction on August 2, went out, probably on its own, three and a half days later, limiting the damage to expensive equipment, and in particular to the four wells that had been drilled before the blow-out.

Source: *The Times of India* (editorial), 18 September 1982

16

Relationship Management – Creating a Collaborative Environment

There have been several measures undertaken by ONGC to assist the local administration in every possible manner including financing projects, providing employment for locals and financial assistance to the schools and colleges. In spite of political agitations in Assam, Tripura and Nagaland, its operations continued to improve, establishing emotional links with the people in the areas of operation. A few instances detailed below would be of interest.

Assam

Assam was having a students' agitation (AASU) at the time I joined ONGC. Within one month of my taking over as chairman, Mr Woodward, a senior member of the Commission informed me on telephone about an agitation and advised me to cancel my visit to Assam, as I was a marked man. I told him that I am arriving there as per schedule. Accordingly I reached the next day and on my landing at the headquarters of the Eastern Region found, as pre-warned, an agitating gathering of people. The executives of the project were too assembled at the Guest House, instead of being in their respective offices for fear of aggravating the agitation. I decided to lead them towards the office, and on the way, one young man approached me who wanted to speak to me. I put my arm around him and took him along to the office where we all listened to what he had to say.

It transpired that the genesis of the problem was the transfer of some non-officers' cadre employees out of Assam to distant locations. At their new places of posting they could neither converse in English nor in Hindi or in the language of the state. Their families were left behind in Assam without support. This was done prior to my joining ONGC, as a sort of punitive measure for their direct or indirect involvement in earlier agitations spearheaded by the AASU, which had resulted in shut downs and dislocation in oil production. At the same time a large number of employees in Class-II and III were employed in Assam from other states creating its own tensions. The student ring leader was personally effected due to the transfer of his father to Andhra. All these problems were handled tactfully to the satisfaction of the employees without watering down the administrative action. It resulted in the distinct improvement of productivity and production of oil.

One day I got a telex message from Mr Mohanta, the Chief Minister, asking me to visit Guwahati to negotiate with the students. I replied that the chairman of ONGC does not negotiate with anyone, but if the students want to meet the management of ONGC, then they would have to come to Delhi, accompanied by a senior official of the state government and the cost of their transportation etc. would be borne by ONGC.

A group of about 10 students accompanied by their advisor Dr Sharma and a very senior IAS officer Bhabesh Chandra Thakurta who later became chief secretary, visited Delhi. Their apprehension that ONGC is not doing much towards the development of the state was set to rest by treating them to a detailed presentation with the help of slides, etc. Developmental activities in Assam were projected in detail and assistance to the society and in particular to educational institutions were made known to them, besides ONGC's contribution to the development of small-scale industry and business in Assam. They were kept busy from 11 a.m to 7 p.m in the same conference room. I was personally present throughout and continued to interact with the visitors. At the end of the meet, Dr Aroon Sharma said, 'Mr Chairman Sir, till this morning we were treated as outsiders, now we stand emotionally attached to ONGC.' A report about the visit of the Sibsagar students (as reproduced below) mentioned about the demands discussed and the way the same were handled by ONGC.

> It is understood that much progress has been made towards fulfilment of 21-point charter of demand of Sibsagar District Students Union, an unit of AASU at the high level meeting of senior officers of ONGC including its Chairman Sri S P Wahi and 10 representatives from District Students Union and AASU held in New Delhi on January. ...

Preference to local youths in matter of appointment in ONGC, adequate compensation to families whose lands have been acquired by ONGC including employing one member of such families in ONGC, cancellation of appointments made during Assam agitation and preference to Assamese suppliers and contractors for all supply and constructions job were some of the demands taken up for discussions in the meeting besides the issues of overall improvement of Sibsagar district's environmental treatment facilities and various problems of ONGC's local employers. ...

Dr Aroon Sarma, Adviser, AASU was appreciative of the attention of ONGC and its positive response.

Source: 'Sibsagar Students Demands Discussed in Delhi', *The Assam Tribune*, 5 January 1987.

Ultimately, the agitators got convinced how the activities of ONGC were going to help the state and national economy. Thereafter, Mr Prafulla Kumar Mohanta, chief minister of Assam had visited Dehradun. The extract reproduced below explains the various activities undertaken by ONGC in the Eastern Region.

The Assam Chief Minister Mr Prafulla Kumar Mohanta visited the Oil & Natural Gas Commission headquarters and expressed his happiness over the activities of the organisation in making the country self-dependent in oil and natural gas. ... He was happy to see the various trades like sewing, knitting, printing, shredding, etc. being done by the dependents of the deceased employees and members of the Samiti (ONGC Mahila Samiti).

At the Institute of Drilling Technology, Mr A.K. Mitra explained to the Chief Minister the various research and development programmes being conducted by the Institute for improving the drilling performance and close monitoring of drilling activities all over the Commission. ... Dr S.P. Wahi received the visitors. At an informal meeting held in the Conference Room he briefed the Chief Minister about the high priority given by ONGC for training and development of human resources.

Source: 'Activities of ONGC in Eastern Region Appreciated', *Reporter*.

These measures were of immense advantage to ONGC and in spite of the ongoing political agitations in Assam, ONGC's production continued to increase from 1 million tonnes in 1980–81 to 3 million tonnes in 1988–89.

The need to maintain an emotional link with the people in areas of operation through informal meets and sincerely planned development activities was obvious from the increase in oil production as the oil production data of Assam revealed. In politically troubled areas the direct involvement of the chief executives and board members is vital, to ensure the proper morale of employees operating in these areas.

Nagaland

On hearing of an agitation to stop oil production, I flew in a single engine helicopter flown by a young air force pilot to meet the Nagaland cabinet. I explained to them the benefits which will accrue to the state by our activities, and I volunteered to pay for the visit of a delegation from Nagaland to visit any part of India or the world to see the economic benefit of oil production activities for the prosperity of the people. After the meeting I was given a ceremonial send off and presented with a hand-woven colourful 'shawl'. The meeting turned the table as the following production profile would show. This also brings out the benefit of managing the local environment with sincerity and mutual respect, and by identifying oneself with the people.

Production Since Inception in Nagaland
CHANGPANG FIELD

Year	Oil (MMT)	Gas (MMSCM)
1980-81	0.0002	0.002
1981-82	0.0236	1.660
1982-83	0.0287	1.663
1983-84	0.0309	2.222
1984-85	0.0369	2.300
1985-86	0.0794	4.160
1986-87	0.1072	5.750
1987-88	0.0865	5.616
1988-89	0.1344	7.250
1989-90	0.1363	8.165
1990-91	0.1210	7.845
1991-92	0.0754	4.848
1992-93	0.0810	10.564
1993-94	0.0839	20.508
1994-95	0.0105	2.373
TOTAL	1.0359	84.926

Source: The years marked in bold indicate my tenure with ONGC (1981/82 to 1989/90)

Production suspended since 11.5.94 as per directive of Nagaland Government.

(Production loss: 250 tpd)

Tripura

I had visited Tripura after being briefed by the geoscientists of the excellent potential of gas in the area. There was only one rig in operation in 1981. In line with the prospective plan up to the year 2004–5, additional rigs were positioned, raising it to 7 rigs in 1988–89. During my meetings with the Chief Minister Mr Nripen Chakravarty, I had established excellent rapport with him. I found the chief minister a simple honest soul who had the vision for development of the state but was short of resources and development plans. I had a number of meetings with him in Tripura as well as in Delhi. Our exploratory efforts resulted in the establishment of large gas reserves in that state. Use of gas for industrial and domestic purposes was proposed including running of vehicles by CNG. A massive promotional drive for the use of gas was launched by ONGC, for power projects, brick kilns, glass, chemical and other industries. Use of gas for power projects was the best option as surplus power could be evacuated to the neighbouring states easily as compared to gas.

The Chief Minister had appreciated the efforts of the ONGC for the development of Tripura and provided us full support for peaceful operations. ONGC continued to get support even from the next government.

Andhra

There was very little activity in Krishna-Godavari and Cauvery basins prior to my joining ONGC. The potential projected by the geoscientists was however very high. I therefore took immediate action to place order on BHEL for rigs and some rigs were moved from other locations. Consequent to these moves, major oil and gas finds were made both onshore and offshore in the area. This opened up huge potentials for finding large quantities of oil and gas.

The production of oil and gas was started from most of the new fields by calling it production testing, to get over the problems of getting clearances through the government. The production was commenced even from Rava (offshore field in Krishna-Godavari Basin) through early production system. This field was gifted away to the private sector in early 90s without even recovering the direct costs, let alone the sunk costs in exploration by ONGC.

Acceleration of our activities and successes achieved in Andhra created the desired image of ONGC. One chief executive of the state public sector undertaking sought meeting with me a number of times. Every time that he met, he would suggest that I should meet the chief minister. Each time I had to tell him that I did not want to bother the chief minister, as I was getting full cooperation. He finally came out with the real issue that the chief minister wanted me to call on him. I told him that the desire of the chief minister is an order for me, and at the appointed time, I reached the chief minister's residence. Chief Minister Mr N.T. Rama Rao and about 10 other senior people, including Mr Bhanu Prasad, former chairman ONGC got up to receive me. After a few pleasantries, the chief minister asked for financial assistance to build a bridge. I readily agreed to fund 50 per cent. Thereafter we moved to the dining hall where the chief minister helped me to eat sweets. This whole incident left a great impact on me about the ability of such a senior man showing so much humility for the sake of the state—a true leadership characteristic. The subsequent successes in Krishna-Godavari and Cauvery basins have confirmed the expectation of our geoscientists both onland and offshore. It is a pity that ONGC is not reaping the benefits of their toil today.

> ...relationship management outside the organisation is as important as within the enterprise.

*****

In the interest of efficient operations of the business enterprise, relationship management outside the organisation is as important as within the enterprise. Based on correct values it helps to create a collaborative environment. The sensitivities of the local people and needs of the administration have to be appreciated and met within limits of corporate governance and social obligations in a transparent manner. A two-way communication needs to be maintained to avoid unpleasant surprises. ONGC continued to have almost trouble-free operations in every part of the country, in spite of occasional political agitations and turbulent environment, during my tenure of over eight years.

Performance Management: Consolidation of the Public Sector Oil Companies

17

Moving through the Bureaucratic 'Jungle'

The Bureaucratic Culture

The creation of a bureaucratic culture may rest mainly with the permanent functionaries in the government, but most of the public sector enterprises suffer from this malady as well, because of its infectious nature. Anyone under the influence of governmental working is bound to be infected in some measure or the other, at some point of time. Even large private sector enterprises within the country and abroad, suffer in some measures with this malady.

Bureaucracy imposes caution in decision-making by highlighting the negative consequences of any decision. Every decision has a risk element, otherwise all decision-making could be taken over by the computers. An average manager would hate to take a decision, unless ordered from the top or pushed from below.

The normal strategy of bureaucracy in the government is to needle and harass the public sector managers till they capitulate to their influence and to their advantage. Bureaucrats in the government have the power but no accountability for the performance of the public sector enterprises except to bask in the glory, if performance is good. They keep their files clean—they call it *Ganga Snan* (dip in the holy water), so that they are never wrong.

Secretaries to the government of India have immense powers of control over the public sector enterprises. Their minions try to impose their influences even more. Public sector executives look like scare crows in front of the bureaucrats. The position is worse if there is a semi-baked technocrat in the government. They are true to the adage *Neem Hakim Khatrae Jan*—little knowledge is dangerous.

One is lucky if the minister and the secretary are working together to a common objective of national interest. It is a big problem if they are working to pursue different objectives and also suffer from false ego of knowledge and position. They however are very sharp in intellect and have excellent powers of communication. Individually most of the IAS officers are brilliant.

I had some experience of dealing with the government and had watched the styles of Late Mr Mantosh Sondhi and Dr V. Krishnamurty, both outstanding managers and successful secretaries in the government. I however lacked their wisdom and maturity in dealing with the bureaucrats and ministers. I have always believed in giving due respect to the chair/position but would be cut and dry if I had strong reasons to doubt the intention or felt any misdemeanour.

Examples of Ministerial Interference

The secretary within a few months of my assuming charge had called for a meeting in his office for a very high value tender for Bombay High platform, against which only one technically acceptable offer had been received from an American company. I had noticed that mostly Western companies had monopolised the offshore work and were exploiting their links in the government, ONGC and World Bank. The secretary expressed great urgency for finalisation of this order. This tenderer already had a major dispute with ONGC on fabrication and installation of two platforms of Bombay offshore. The action of the secretary to force the decision on this case when no proposal had been submitted by the ONGC was unusual. I had however disagreed and the case was re-tendered and there was a saving of over Rs 100 crores. I had also noticed on joining that the specifications for the tenders were being prepared to suit the capabilities of only a few potential tenderers—understandably only from the Western world.

Bureaucratic Needling

In line with the nature and style of bureaucratic functioning, unnecessary correspondence started from the Department of Petroleum to which I was forced to respond suitably. The D.O.

letter dated 30.12.1982 sent to joint secretary to the Government of India, with copy to secretary (petroleum), has been appended below.

"Dear Shri Khanna,

I have before me two letters from you dated 7.12.1982; one expressing your distress on the speed of establishment of a base at Kakinada and the other on drilling targets.

I am deeply pained and concerned by the contents, tone and tenor of the communications.

I always believe that a soft answer turneth away wrath; therefore, I have been exercising very great restraint and politeness in replying these communications of the Ministry couched as they are in a very aggressive language.

I have a feeling that of late while the Ministry is on one hand talking about autonomy, changing the ONGC Act to give wider powers to the Commission, greater delegation of powers and decentralization, etc. on the other hand, the Ministry is actually wanting to hold ONGC on a tight rein, almost on a leash.

The need for power seems to make for the psychological compulsions to magnify the need for review and control. I believe that a sound management principle is that persons who have been given the responsibility must not only be given the authority but should also not be hustled, harassed and chased except where the principle of management by exception holds. Review and control cannot be so detailed that initiative is stifled, freedom to perform is throttled, and flexibility in approach subordinated to regimentation.

These letters compel me to feel that ONGC is being treated as a subordinate office of the Department of Petroleum; that the organization which is multifaceted, complex, growing at a tremendous pace, managed by highly competent professionals is being subjected to frequent needless needling which is adversely affecting the morale of the organization. (Information about drilling targets was given in the next few pages.)

Regarding your letter on the establishment of base at Kakinada: it involves land acquisition, infrastructural facilities, buildings, etc.

We have already initiated action for land acquisition. There is not even an air-strip at Kakinada. I really wonder if there is any realistic appreciation of the time required to establish the base at Kakinada and the matter has been considered in all its aspects before this letter was issued. We are committed to establish Kakinada Base and every possible action will be taken.

This letter was issued even before my confirmation to the post of Chairman, ONGC. Thereafter I was subjected to a very amusing drama which is narrated below.

Amusing Drama

Late Mr Gargi Shankar Misra was appointed as minister of state in the ministry of petroleum. I was away on tour to Assam on the date of his joining. I asked my private secretary to get an appointment with the minister. I was told that he was already looking for me. On arrival in his chamber I was well received by the minister. He started the conversation saying, "I am also an Army man. I have killed many people. I allow allegiance only to God or Indira Gandhi." He kept on repeating such irrelevant remarks, till his assistant private secretary brought a piece of paper, which he passed on to the private secretary to the petroleum minister (a brilliant IAS officer) who was present there, possibly to witness the drama. At that time, in a formal meeting with the minister, I was a little surprised by his presence. He returned the paper with a remark that it was very strong. The minister thumped the table and said that he was also very strong. He then passed on the paper to me and said that it was his love letter to me.

I was amazed at the way the things had unfolded. After perusal of the contents of the letter, I braced myself and decided on a counter attack and said, "Sir, you have the right to issue this letter, but you have listened only to the story of the administrative wing headed by the secretary, but not to the commercial wing headed by me. I am not only the seniormost chief executive in the petroleum sector, but also the whole public sector. I have the privilege of representing against this to the prime minister." I got up and wanted to leave. "You have got angry for nothing. I have not even signed it." He took the paper from me, tore and threw it in the waste paper basket. Through this process both sides conveyed to each other the power they wielded. This minister never called me again to his office.

For the rest of my tenure there was total peace except for two experiences with two secretaries who joined later. However with my upright approach, I was free from long drawn meetings in the ministry and with less interference we were able to perform even better than international norms with the result that ONGC continued to receive international recognition.

Performance Hurdles—Bureaucratic Ego

There were outstanding executives at ONGC who were able to develop relationships at the highest level at different ministries to get financial, environmental and administrative clearances. One of the joint secretaries went up to his secretary and complained that ONGC was directly dealing with other ministries and getting the approvals. This secretary was very matured and advised the joint secretary to take credit for whatever ONGC was getting done, as the final orders had to be passed through the petroleum ministry. ONGC had to literally carry the files from one desk to the other to get the approvals to meet their performance targets. In fact, in the case of wage revision of the employees, I had to personally get hold of the file from the Bureau of Public Enterprises and get approval from the expenditure secretary in the ministry of finance.

The next secretary was an extremely hardworking intellectual with wide experience in different ministries, as chief secretary for a short tenure in his parent state and a tenure abroad in an international agency. He was in the habit of interacting with a very ambitious senior member of my team and write letters on issues on which he had little knowledge and were not the areas of his concern. The member concerned had to be cautioned and the secretary had to be suitably replied. He was very fond of publicity. Suddenly one morning *Economic Times* brought out a news item 'Shake up in ONGC through Reorganisation ordered by Government'. This news item was based on the handout issued by the ministry.

There was a lack of understanding of a matrix organisation structure which had been introduced in 1984 with government approval. We had put up a proposal to the government for appointment of executive/regional directors in the same scale of pay as member (board level). This was essential to give desired powers at the regional level to ensure accountability for quantifiable results. Approval of this proposal was being projected as reorganisation ordered by the ministry.

The confusion created by the media based on the handout of the government had to be removed by advising all the Regions of ONGC not to take note of what had been published. A D.O. letter had to be sent to the secretary requesting him to issue a rejoinder to the press. This resulted in a lot of infructuous correspondence. No changes to the existing organisation structure were envisaged and nothing was done. To satisfy the ego of the secretary, member (operations) was redesignated as member (natural gas). This designation remained only for a month, as change in designation was meaningless.

The organisation structure of ONGC and project clearance was the subject of examination by COPU (Committee of Public Undertakings) (1985-86) and duly recorded with recommendations in the eighth report presented to the Lok Sabha on 28 April 1986. The action taken report (Seventh report of COPU (1986–87) was presented to the Lok Sabha on 25 March 1987. We were forced to state the facts to the COPU, during our examination, which were at variance with what had been presented by the secretary petroleum. Chairman COPU was obliged to ask for the government file, which had contained exchange of correspondence with ONGC. A few extracts from the reports are reproduced below.

> The Committee finds that a reorganisation scheme which seeks to fully implement the concept of centralised policy making and decentralised administration was introduced in the Commission in July 1984. This scheme was introduced after carrying out a SWOT analysis by ONGC. According, to ONGC the reorganisational structure which came into operation from July 1984 had positive impact on the working of the Commission and had already started giving desired results and the operational efficiency was on the increase in every area. The Ministry of Petroleum had strangely enough a different assessment of the scheme. It has been stated by the Ministry that after one and a half years of this reorganisation, the Government reviewed the position and found several weaknesses in the system. According to Ministry it was found that the different functional Groups had not yet been able to organise themselves as Business Groups acting as cost and profit centres and the ONGC had been functioning essentially as a centralised unit with a common budget. A new reorganisation scheme which was to be effective from 1 April 1986 was being introduced to bring about improvement in the changes already made.

The Committee finds that in fact the scheme of reorganisation introduced by ONGC is sought to be improved although Ministry have claimed that it was a new reorganisation scheme. On the one hand, the Secretary of Ministry deposed before the Committee that too frequent changes should be avoided, on the other hand Ministry themselves are bringing about changes within one and a half year of the introduction of reorganisation by ONGC. The Committee are not able to appreciate this situation. The Committee hope that Ministry had discussed the changes in reorganisation with the ONGC before introducing them. The Committee will like the Ministry to clarify this and inform the Committee after six months of the results achieved by the new reorganised set-up.

Source: Eighth Report of COPU, 1985-86

*****

Reorganisation of ONGC

The Committee had noted that there were frequent changes in organisational structure of ONGC in the past, viz. in 1974, 1976, 1978, 1981 and 1984. Another reorganisation on scheme was sought to be introduced w.e.f 1 April 1986. The Committee wanted to know from the Ministry whether the reorganisation scheme of 1986 was a new scheme or it was only an improved version of the 1984 reorganisation scheme. The Committee had also desired to be informed about the results achieved by the introduction of the new reorganised set-up.

In their reply the Government has merely stated that to ensure better operational performance, greater coordination amongst various functional groups and accountability for overall performance in defined areas, six regional centres have been created each under the control of a Regional Director, with overall responsibility for the operations and result in each region.

The Committee regrets to note that in their action taken note, Government has only given the details of the reorganisation scheme, which are already known to the Committee. To what extent the new reorganisation scheme introduced with effect from 1 April 1986 represented an improvement over the erstwhile set-up and whether the change in the reorganisational set-up had been discussed with ONGC before introducing them have not been clarified. The Committee wishes that replies to their observations should be complete and expressed in unambiguous terms. The Committee will await necessary clarifications from the Ministry.

Source: Action Taken Report of COPU, 1986-89

Bureaucratic Accountability

There was no reorganisation scheme introduced from 1 April 1986. Instead of admitting the confusion which had been created through the media, the government had no option but to give an ambiguous reply. Bureaucrats get away with all the confusion and have no accountability for the damage they create to satisfy their ego.

During the last phase of my tenure with ONGC I had to deal with a secretary who was very sophisticated in his behaviour. The secretary was looking forward to my early retirement so that he could indulge in activities prejudicial to the interest of ONGC to satisfy his ego or otherwise. I came to know much later that the secretary through a written note had informed the cabinet secretary that I was helping Mr V.P. Singh (ex prime minister) in his nefarious activities against the government of the day. The fact is that I had never met Mr V.P. Singh on one-to-one basis even in his official capacity when he was a minister in the Cabinet. When the chips were down the secretary had to eat his humble pie.

I had made it known to the minister Shri Braham Dutt, my desire to retire on 6 November 1989, but he had persuaded me to stay. I was however suddenly relieved from my position through a telephonic advice one evening, as soon as Mr V.P. Singh had taken over as prime minister. This would indicate my relationship with him.

Honest Compulsions of Ministers

During my tenure as Chairman ONGC, I was receiving chits/letters/employment requests from one of the ministers. I went up to the minister and explained to him the system and procedures being followed. He was an honest good soul and mentioned that being a public man he would receive and would send such papers to me. "You may deal with those in line with ONGC procedures. In case any work gets done in the normal course, then please advise us," he added. This was a reasonable request.

We did receive a few genuine complaints about delayed payments and playing around with tender specifications and terms. The people concerned were cautioned or shifted from their seats or stations, to give a message all round. In most of these cases personal integrity of the executives was not involved but interpretation of numerous instructions did create ambiguity and lack of clarity.

Corporate Governance

Simply stated, corporate governance is effective management of resources and constraints to generate financial surplus for further growth. It means managerial effectiveness and leadership with ability and personal courage. In this process, no individual or group should take undue advantage for personal gain at the cost of the interest of the enterprise. It entails competence and integrity not only on the part of professional managers of the enterprise but also on those entrusted with regulatory and controlling responsibilities.

> Simply stated, corporate governance is effective management of resources ...

The sine qua non of corporate governance is that all those who are involved in it must be driven by a common objective, i.e. to promote the interest of the enterprise. Important factors for ensuring effective management of resources are organisational structure, technology, management practices in line with emerging business environment, managerial effectiveness, leadership abilities to plan for the future and care for the morale and motivation of human resources.

Present State of Governance of Public Sector Enterprises (PSEs)

A recital of the characteristics of good corporate governance will reveal how far removed we are from the given minimum acceptable standards. At the very outset the bureaucrats performing the so-called regulatory and ownership role in the ministries scarcely ever share a common objective with the PSEs. They see their roles more as fault finders and prosecutors. Even if they were members on the board of directors, they would never sit with the PSEs to formulate a proposal on common understanding with the view to carry the enterprise forward. They prefer the proposals to be put up to them and then they will sit in judgement leading to interminable exchange of notes in acrimony and frustration.

It is our country's misfortune that our administrative systems continue to be the inheritors of the colonial style. By definition the bureaucrats thriving in this system are self-preservers and operate on a distrust matrix. **They feel secure only when an external**

consultant holds their hands, preferably from the Western world. The in-house advice, however competent, is held with suspicion. They dislike risk in decision-making. Their careers do not depend upon the success or otherwise of the PSEs under their control, but only on how well they keep themselves out of trouble. It is amazing (and it is certainly a matter of self-congratulation for the successful PSEs) how in the face of the obstacles placed before them, the PSEs have been able to achieve so much.

It is a matter of historical record that leading PSEs in the areas of oil, power, engineering, atomic energy, etc. have come up to play their stellar parts not so much on account of the bureaucratic faith and support to them as on account of the inspiration bestowed on them by the political leadership at the highest level at the respective times. As an illustration, but for the vision and inspiration of the post-Independence prime ministers, there would have been no oil discoveries in the country. The bureaucrats never believed that India would find oil within its territory because the Americans said so! For 15 years the bureaucrats treated the ONGC as a temporary organisation and did not confirm any of its officers to speak of any career advancement for them.

The irony is that the bureaucrats belonging to the several civil services are drawn from amongst the best available talent in the country. It is only the best amongst them who rise to major controlling and policy-making echelons in the hierarchy. Many of them are individually bright and brilliant. It is the inherited administrative system that sucks these bright and brilliant men and women into a cesspool of involuted thinking, suspicion, disbelief and distrust and, not the least, timidity. The system stymies them into a state of perpetual fear and further into more abhorrent fear of being held individually accountable.

The system does not accept a bureaucrat saying that he believes in the PSE under his control. Even if he personally does, the system will want him to say that he is 'led' to believe. The bureaucrat seldom 'knows' on record. The 'knowledge' is always brought to 'his' notice. A typical bureaucrat will never be caught initiating a proposal in favour of a PSE. He will only receive a proposal. Even then he does not study the proposal. He will have it examined by his juniors and then consider it. The colonial genome ensures that someone else is accountable and never the bureaucrat.

Regrettably 'Blame Game' is our national hobby. Witness our several legislative bodies spending unconscionable time more in finding fault with each other (particularly the present with the past) than on sitting together, as constitutionally mandated, in constructive debate to take the country forward. What is often overlooked is the eternal truth that in the business of the PSEs, enduring and successful decision-making is borne out of a collegiate approach based on mutual respect and trust and not on a divisive approach as is in practice today.

> Management of business involves perspective planning, quick decisions and risk taking.

Ownership role today rests with such bureaucratic administrators in the ministries controlling their PSEs, with accountability to the parliament being seemingly perceived through ministerial supervision. Management of business involves perspective planning, quick decisions and risk taking. The bureaucratic administrators in the ministries have short tenures and tend to take only short-term views virtually co-terminus with their respective tenures, the objective being that their yardarm should always be clear. The colonial system psyches them into being risk averse. Short-term views will not help the long-term interest of the business of the PSEs. The bureaucrats do not have adequate knowledge of the complexities of various business and technologies. Even the basic principles of management are generally unfamiliar to them.

The objectives and conditions of service of the bureaucrats in the ministries and the managers in the PSEs are not the same. Those in the ministry also suffer from superiority complex, as appointments of the top managers, their service conditions and matters like visits abroad are controlled by the ministry. All these lead to a subservient culture in most of the PSEs. The chairman and the members of the board are summoned to the meetings by even relatively junior bureaucrats in the ministry and made to wait for long hours beyond appointed times. Those managers who stand up to these unjustified demands of these functionaries in the ministries and urge quick action on their part are subjected to different types of harassment, sometimes stretching years after their retirement from the PSEs.

Not that there have not been exceptions to this general run of bureaucrats. Fortunately for our country, there have been a few notable and very creditable exceptions. Also there has been quite a few top level professional managers who in their passion and loyalty for their profession and more importantly in a gallant spirit of patriotism for their new and resurgent country have performed for their PSEs, in virtual defiance of the bureaucracy and not infrequently at great risk to their personal careers. It is because of their collaboration with the highest level of political leadership that the country is at least where it is today. **If only these few were many, we would have banished poverty decades earlier and achieved prosperity levels that would be the envy of several developed countries.**

Let me now take you through my experiences and continuous efforts, even till today for consolidation of the public sector oil companies, to stand the ruthless competition in the present globalisation environment.

18

Re-engineering the Oil Sector—Globalisation Imperative

Public sector oil companies have played a pioneering role in putting India on the world oil map. As mentioned in an earlier chapter the history of oil exploration in India dates back to 1867, only eight years after the first oil well was drilled by Col. Drake in USA. During the British rule the whole country was available to the British private oil companies with their financial and technological strength to explore and produce more oil. However till 1947 only 0.25 million metric tonnes of oil per year was being produced, and oil was discovered by them only in Assam. It was only after Independence and the nationalisation of the oil sector that major oil discoveries were made by ONGC in Assam and other parts of the country like Gujarat, Andhra Pradesh, Tamil Nadu and more particularly, in Bombay offshore.

Since the beginning of the 90s (I quitted ONGC in December 1989) there has been a campaign to denigrate the public sector ONGC by gifting away a large number of oil fields discovered by it to the private sector. And to make it a real gift, it was rather presented to them without recovering even the direct costs let alone the sunk costs over the years on exploration activities all over the country. *The fact however is that oil production in the country has gone down after 1989-90 and at best has been stagnant.* Exploitation of the sedimentary basins that were made prospective by the national oil companies (mainly ONGC) have been exploited by the private oil companies (national and international), but not a single oil/gas strike has been made by them in the 20 virgin basins.

ONGC till privatisation of the oil sector was getting Rs 203.41 for onshore oil and Rs 331.65 for offshore oil per tonne up to 10

July 1981 and then changed to Rs 1021.00 per tonnes in 11 July 1981, both for onshore and offshore oil. Furthermore, in spite of repeated requests, no increase in the price was allowed that could have generated enough finance for ONGC to exploit the marginal oil fields. ONGC had all the technological resources to accelerate activities in the areas which had been made prospective through their own exploratory efforts. They had the ability and credibility to borrow money to exploit marginal fields if they had been given even near international oil prices, which were given to the private sector after gifting them away the discovered oil fields of national oil companies. This opened the floodgates to the private sector to exploit the sedimentary basins made prospective by the national oil companies. The comptroller and auditor general (CAG) of the Government of India severely indicted the government on this deal and even the Cross Party Estimates Committee criticised the deal in its report of 1998–99 and 1999–2000. The CBI had initiated an enquiry and its initial investigations suggested a near sell-out (*Business Standard,* June 1996, *Times of India,*1 November 2000). These however had no effect except to the extent that medium-sized discovered fields were similarly not gifted away any more. This action would prove to be a historical blunder either through ignorance or design by people for self-interest. Historians, I am sure, will find the truth.

A few news items which appeared in the 90s, covering the shape of things to come, in an attempt to demoralise the public sector oil companies in the field of exploration and production are appended below.

Government to Hand Over Oilfield to MNCs?

Is the Government planning to hand over the biggest unexploited offshore oilfield, Neelam, to multinational companies (MNCs) for development? This question is being asked in the oil industry circles with the Government's decision to accord low priority to the development of this field which has the potential to produce at the rate of 6 million tonnes per annum.

At one stage Neelam project was given the highest priority as it was crucial to meeting the 8th Plan target. The project was tendered and re-tendered without a decision for the last three years. Finally, the Government took a decision on January 31 to issue a letter of intent to

Hyundai of South Korea for turn-key execution of the project on the condition that the contractor would arrange credit from its EXIM Bank.

All the four offshore projects in Bombay have gone to South Korean companies with Hyundai getting two and one each to Daewoo and Samsung. In the list of priority submitted to the Korean EXIM Bank, India's economic minister in the embassy in Tokyo put Neelam as the last.

> ! Sources say there is a deliberate attempt to make out that the oil scene is in a mess. !

This created an impression in Korean EXIM Bank that Neelam was not important in the Government's scheme. Accordingly, it offered an export credit of $363 million against three projects of ONGC. The projects are SHG process platform, PQSU and ABCDE well platforms. In addition, a credit of $141 million was offered for vessels to be purchased by Shipping Corporation of India (SCI).

ONGC needs at least $1.173 billion to finance these projects. There are strong indications that the Korean EXIM Bank, may not accommodate the full foreign exchange requirement of Neelam which is around $400 million. It may provide half of it if the Government takes up the issue at the political level.

The sudden loss of interest in Neelam has triggered speculation that the project may be handed over to MNCs. There is a powerful lobby both within ONGC and the Government to privatise almost everything in the oil sector. It is alleged that some people are doing this to secure jobs in the World Bank after retirement.

Some time back, a senior executive in ONGC mooted a proposal to hand over Ravva field in Godavari and Gandhar Phase III to private parties for development. This was opposed by Mr P.K.Chandra, vice chairman, who has since retired. He argued that ONGC would still make a profit even if it developed these fields with money borrowed from the commercial market. ONGC enjoys better credit rating than any private company in India.

Sources say there is a deliberate attempt to make out that the oil scene is in a mess. A managerial problem at the top level is sought to be twisted to prepare the ground for inviting foreign companies into this area. The Government has already received proposals from oil giants like Shell, Chevron and Total offering to participate in the fourth round of bidding if they are allowed to develop proven offshore fields. A decision on these offers is expected after April 15.

Sources say the oil scene is witnessing a spectacle of public sector executives, top bureaucrats and even ministers vying with each other to project themselves as the advocates of privatisation. This, they say, is prompted by the belief that success in life lies in identifying with the multinational lobby.

Source: R. Sasankan

*****

ONGC Probe

Mr A. B. Vajpayee today demanded a high-level inquiry into the working of the Oil and Natural Gas Commission, which he alleged, was 'largely responsible for the current BOP (balance of payments) crisis'.

In a letter to the Prime Minister, Mr P.V. Narasimha Rao, he said the ONGC's 'deteriorating performance in the first two years of the Eighth Plan had an adverse effect on the BOP position. While production surpassed target by 0.39 million tons during 1989-90, by 1990-91 there was a shortfall of 3.7 million tons. During 1991-92 production again fell to 28 tons against the target of 34.7 tons,' he said.

'It is intriguing that a premier public sector company like the ONGC, which has made the country proud in the past, is being transformed into a sick unit,' the BJP leader said.

'The recent agitation by ONGC's officers and staff had brought into focus the bureaucratic style of functioning of the present management and the lack of dynamism and leadership qualities, which are required to manage it,' he added

Source: Statesman, 30 September 1991.

*****

Known Oilfields to be Thrown Open to Foreign, Indian Firms

The Government has decided to let private and foreign companies develop discovered oil and gas fields. The finance ministry has given its consent to the proposal, which is now being sent to the Cabinet Committee on Political Affairs (CCPA).

This move, a major departure from the existing policy which allows only public sector companies to develop discovered fields, is expected to attract large amounts of foreign equity. Private sector participation will be sought for developing medium- and small-sized fields. Public sector companies will continue to develop the large fields.

The ministry of petroleum and natural gas is formulating the terms of agreement which would govern foreign and private sector investment. Since these fields are already discovered, the Government is likely to impose certain statutory levies.

The offer for development by domestic or foreign companies would be on the basis of contract, lease, joint venture or sale, sources said.

Development of these fields will run parallel to exploratory efforts by foreign and Indian companies in the fourth round of oil bidding. While in the fourth round the risk venture capital will be involved, the development of discovered fields will involve no such risk. This fact is inducing the Government to impose certain statutory levies.

The ministry, in consultation with the public sector oil companies, is currently identifying fields to be thrown open to the private sector. According to a quick estimate, there are about four or five medium- sized fields and over 15 small isolated small pools scattered all over the country.

The idea for allowing private or foreign companies to develop discovered fields was mooted following resource crunch facing the Oil and Natural Gas Commission (ONGC). For executing all its projects, ONGC will require roughly $4 billion in foreign exchange. Development of these fields would have meant outgo of more exchange or increase in foreign debt.

Source: *Economic Times,* 18 March 1992. Also see Appendix B

Handing over Panna-Mukta to JV a dead loss

The Government has suffered a loss by allowing development of Panna and Mukta fields under the joint venture arrangement instead of ONGC, on the terms laid down in the Production Sharing Contract (PSC), the report prepared by the Indian Audit and Accounts Department has said.

This has been alleged repeatedly in the past two years, but the audit report has made a detailed analysis for the first time.

The report questions, one, the comparative economics on the basis of which the fields were taken away from ONGC and handed over to the JV, that is, the Reliance-Enron combine with ONGC holding 40 per cent stake. Two, it says that the actual operating costs by the JV in

December 1994-December 1995 were much higher than the cost estimates submitted in the bid by the operator (Enron).

The Ministry finally accepted ONGC's earlier cost of development (Rs 3,367 crore with Rs 1,897 crore towards development of Mukta), but lowered the recoverable reserves estimate to 14.57 million tonnes (4.57 million tonnes for Mukta).

The report points out that it is not correct to carry out the comparative analysis taking the revised (lower) reserves of 4.57 million tonnes at the same development cost (Rs 1,897 crore), which was meant for recovery of much higher reserves from a larger area.

NPV of government take from Panna-Mukta under PSC worked out by government was much higher (Rs 2,230 crore) than that worked out by ONGC (Rs 1,809 crore).

The report says: **This policy ... "may result in NOCs fighting shy of taking risky exploration initiatives because of a constant fear of handing over a discovered field to private parties on a platter when it was ready for yielding profits."**

Source: Ruenna Burman, Economic Times, 5 July 1996

The final deduction of the Indian Audit and Accounts Department report is unfortunately collaborated by the downward performance of the National Oil Exploration companies, as a result of privatisation policy of the Government. There is continuous outflow of competent oil-professionals from the public to private sector oil enterprises. **We have not even learnt from the experience of Russia who have again started nationalising the oil sector.**

Shortly after taking over as chairman ONGC, my interaction with a large number of national and international specialists, led me to conclude that certain major transformations were required in the oil sector, if energy security is to be assured to the country in the 21st century. Some of these major conclusions and subsequent actions have been detailed here.

- An integrated National Energy Policy was to be evolved with a long-term perspective in mind. In ONGC at least, a 20-year perspective plan was developed.
- The explosive rate of increase of consumption of oil (the rate being higher than the rate of GDP growth of the country) and the slow progress in tapping alternate sources of renewable energy made me realise that 'self-sufficiency' in oil may be a

difficult proposition for the ONGC to meet from within the country alone. It was therefore necessary to strengthen the overseas arm of ONGC's Hydrocarbons India and rechristen it as 'ONGC Videsh' to follow an aggressive policy for acquiring acreages abroad, both for exploration and development. With this in view the exploration blocks in Vietnam were the first for us to get in the basket.

> We have not even learnt from the experience of Russia who have again started nationalising the oil sector.

- Synergy for energy and an aggressive bidding for global tenders would require Indian oil companies to be of comparable size to the major foreign oil companies in their market capitalisation and turnover. Therefore, a vertical integration of the Indian public sector oil enterprises was needed.
- With the world trend changing towards more and more major gas discoveries whether in Russia or in the Middle East, it was considered prudent to deliberately chart out a path to move towards gas for the 21st century.
- Since the areas of occurrence of large gas deposits and the areas of their peak consumption may not always be the same, it was felt necessary to connect the country by a 'National Gas Grid Pipe Line'. A conceptual plan was prepared by ONGC and submitted to the Government.
- ONGC studies indicated that there was thick coal seams occurring between the oil-bearing sands in many areas of operation, at depths below the usual mining capability. In north Gujarat alone, these coal deposits were more than the entire proven coal reserves of the country as a whole. ONGC took up the challenge for underground coal gasification so as to exploit these reserves at the opportune time.
- In view of the commitment towards protection of the environment, India had to fall in line with some of the developed countries in the use of clean fuels. Accordingly ONGC started R&D efforts with some of the transport companies like Gujarat State Road Transport Corporation and Ashok Leyland in Chennai for use of compressed natural gas (CNG) as fuel and ONGC was the first organisation in the country to use CNG for its own transport in Tripura.

Trials and Tribulations – Re-engineering the Oil Sector

The oil sector was nationalised after Independence in view of the correct vision of political leaders of the time. There was lukewarm support from the Western developed countries for India to develop as a vibrant oil sector. Former Soviet model was accepted for oil sector development in view of the willing support from USSR and former Eastern Block countries. As a result almost 20 different enterprises got established under governmental bureaucratic control without the expertise, similar to the one with Soviet bureaucrats. **It is unfortunate that the corporate governance of the public sector in our country is lopsided resulting in serious constraints in the optimal utilisation of resources.** In this context I narrate here an incident.

One day a D.O. letter was received from a joint secretary to hand over ONGC gas activities to Gas Authority. The letter was consigned to the file FF (File and Forget) as no prior discussions had taken place on the subject and had serious safety and commercial implications. No action was taken as long as I was the chairman, in spite of a lot of pressure from the government and the Gas Authority. In fact, Gas Authority should have been established as a subsidiary company of ONGC. (Appendix R).

The whole world is consolidating but we continue to disintegrate so that our bureaucrats and politicians can divide and rule. **We need to re-engineer the oil sector so that ownership role is passed on to professionals, who are subjected to the laws of the land and additional controls through necessary amendments to the company law, can be ensured.**

No owner in his senses will allow his own company to fritter away the resources in multiplying individual efforts and fighting each other in the national and international arena. This is what is happening through default or by design to kill the goose which has been laying golden eggs. There is a need to integrate all the Oil PSEs and bring them under one professional umbrella. This will ensure optimum utilisation of resource to face the stiff competition from the multinationals. We have many models (Premix, Stat Oil, Exxon, etc.) to adapt to our needs.

Late Shri Rajiv Gandhi, the prime minister had admired the performance of the oil sector in general and ONGC in particular. He always exhorted the public sector for more dynamism. This can happen only if the public sector is restructured. (Appendix-T).

Synergy for Energy

The government had set up a committee of very eminent people (mostly bureaucrats) to address the issue and integration of PSOCs (public sector oil companies). The recommendations in the report if implemented however would further deteriorate the working of the oil companies. The fundamental principle is that management of operating companies and those who play the ownership role must have similar conditions of service and share the operating risks. The need to adopt a foreign model operating under different political and administrative system was not appropriate. We had and still have many successful models within the country and abroad in countries like Italy (ENI) and Mexico (Pemex).

> There is an imperative need to develop an Integrated National Energy Plan with the help of technocrats...

The recommendations of this committee were steered by two very articulate but highly self-opinionated former bureaucrats. The government had rightly not accepted the recommendations. However, any further delay in consolidation of PSCs will continue to result in sub-optimisation of resources, bringing to focus inefficiencies of PSEs in this highly competitive environment, when MNCs are integrating through mergers and acquisitions.

Move Away from Dependence on Oil

Oil is too vital for the future economic development and its management cannot be handled as for other commodities and resources. For energy security, there is, however, an urgent need to move away from this depletable resource through demand management, and develop alternative sources of energy. There is an imperative need to develop an Integrated National Energy Plan with the help of technocrats with the sole objective of moving away from total dependence on oil.

Reproduced in the following pages are extracts from my letters to different authorities and talks and address at different occasions to show how painstakingly, I had been pursuing the ideas expressed earlier.

The following extract is from a letter addressed to Dr Arjun Sengupta.

19 December 1984

... As desired some of the important points made by me in the 10 December meeting are detailed below:

Organisational structure and linkage with ministry

There are more than 209 Public Sector Enterprises and operating almost in every sector of economy. In the same sector of the economy, more than one PSES are operating. Their working is being coordinated by the respective administrative Ministries, rather ineffectively due to different objectives of the Administrators and Public Sector Managers. The role of the administrative Ministry is to 'Administer' whereas the role of the corporate enterprise is to 'Manage'. Public Sector has to work on commercial lines which involve speed in decision-making, calculated risk taking and innovative approach for achieving quantifiable results.

The public sector enterprises in the same sector should be controlled by a holding company and the Chairman of the holding company should report to the Minister directly and should also be designated as Secretary to the Government. The Board of the holding company could comprise of Administrators, Members of Parliament and other professionals from industry. Each subsidiary company should have an operating board comprising exclusively of executives from the company.

The holding company would be responsible for laying down policies, approval of annual and Five Year Plans, act as an Empowered Committee for the approval of the projects, review performance of the subsidiary companies and look into the growth of the industry by formulating perspective plans. The Chairman of the holding company would interact with the statutory bodies, a function which presently is being performed by the Secretary of the Administrative Ministries. This organizational structure would insulate the Public Sector Enterprises (subsidiary companies) from the various governmental agencies and the Parliament, thus allowing them to concentrate on operations.

The concept of the holding company will ensure vertical as well as horizontal integration of operations in each sector of economy, resulting in optimization of resources.

Project Approvals

Public Sector Enterprise should work on the basis of long-term perspective plans which will ensure the time frame required for decision-making and the establishment of long lead infrastructure and industrial facilities in the country. This would enable development of indigenous industry as long-term needs of the Public Sector would enable correct investment decision in time to meet the needs. The long-term plan of the ONGC can be quoted as an example which has generated the right interest and enthusiasm in the Industrial Sector to invest in the oil sector.

Once the Planning Commission has approved the Capital Budget, no further checks in the form of administrative or financial clearance should be required. The approval of the Capital Budget is a result of detailed in-depth, series of discussions with the Planning Commission followed by discussions at the highest level. The present system of Project clearance through a large number of agencies is time-consuming without any advantages. The Board of Directors of the Holding Company should act as the Empowered Committee.

The oil sector has been and will continue to play an important role in the economic development of the country. It has to grow with speed which can be ensured by removing the constraints inhibiting its growth. The main constraint is the present system of Project approvals and clearance of purchase proposals. 75% of the sedimentary basins have still to be explored with speed and urgency. ONGC should, therefore, enjoy financial powers commensurate with its needs.

Technology Development

To take advantage of the fast changing technological developments in the world, the chief executives and board members should have the necessary freedom to interact with the international environment. They should have the freedom to move abroad to respond to their counterparts with speed. Tremendous amount of freedom is enjoyed by large companies in the world of the stature of ONGC on such matters.

Modernisation

Electronics and modern e-communication will play a major role towards improving the efficiency of Industrial Enterprises. Technology obsolescence in this area takes place in 3 to 5 years and therefore the need for quick decisions for import of technology and equipment. The present system of working through DOE is time-consuming and frustrating. The Department of Electronics therefore require to be restructured to respond to the needs of the Industry.

Subsidies

Public Sector should not be asked to provide subsidy to other Public Sector (price preference) or to the indigenous industry. This should be done if required by setting aside a separate fund in each sector of the economy.

Some issues have been further projected in the article titled 'Petroleum's Past, Present and Future' (Appendix I), based on the author's address on 16 December 1988. An article, dated 16 September 1990 which covers the futuristic thoughts for energy security for economic growth is provided as Appendix C. An extract from an article 'Autonomy and Accountability in Oil and Gas Sector' is presented below.

Integrated Approach

Natural gas development can take place efficiently only if there is an integrated approach between the gas producers, transporters/distributors and markets, a role which can only be played by a commercial organization and not by the government. No doubt support from the government would always be required.

This calls for a major restructuring of the public sector, and in particular, the oil sector to make sure that the companies are managed by professionals, and have only an arms length relation with the government. There are many good examples available in the world to follow.

Perhaps India is the only example where oil sector has been disintegrated into 16 parts and each part is controlled by different functionaries of the government resulting in lack of coordination and unbalanced growth. The oil sector has suffered because of this organizational constraint, as commercial/business initiatives are not possible under the present system. Otherwise, there is no reason why oil-refining capacity is not at least 30% more than what it is today and oil exploration and production companies are not operating in many basins abroad and have not acquired some 'High Technology Companies' in the oil sector abroad during recession in the 1980s.

There is a need to ensure integration of the oil sector either through a holding company or supervisory board. The performance of oil sector can further improve and can help in accelerating economic development.

Source: *Urja Oil and Gas International*, January 1992

The extract given below details the need for re-engineering the oil sector, and is taken from a presentation made through a discussion paper in a seminar organised by ASSOCHAM in Delhi, on 23rd November 1998.

Management of Oil Sector

... the bureaucrats, in the Ministry of Petroleum and Natural Gas, who are individually brilliant but the system does not allow the decision-making with speed and **business like manner.** The Ministry has to depend entirely on expertise from outside and does fall prey, sometimes, to the advice of people who have either vested personal or business group interests. ...
There is not a single **integrated** (upstream and downstream) oil company, to face the competition from the International Oil companies in the present liberalized global environment. **There is inadequate coordination between the enterprises/organizations under the control of the Ministry of Petroleum and Natural Gas, resulting in multiplicity of efforts on the same subject, resulting in infructuous expenditure and wastage of National Resources.**

In the present highly competitive environment, the International Oil companies, all over the world are reducing costs through **strategic** alliances and in some cases mergers. Our individual enterprises are going their own way, without positive inputs from the coordinating agency. In the absence of a 10-20 years perspective plan and vision for the Oil Sector the individual enterprises will grow in an unbalanced manner. This will have serious adverse impact on the economic growth.

The Oil is being used as a 'swing fuel' and demands are being projected on the basis of easy availability of Oil. This unrealistic approach will create serious oil supply problems in future. There is an urgent need for the preparation of an Integrated Energy plan with an objective to move away from dependence on oil. To ensure efficient management of this critical limited resource, major programmes for conservation and switching over to alternative duels will have to be launched.

Re-engineering of the Oil Sector

In view of the foregoing, there is an urgent need to restructure the **Management** of the Oil Sector, to ensure optimum efficiency. The creation of a few integrated Oil companies through mergers/ strategic alliances of existing E&P companies with refineries will have to be considered. These issues should be discussed in a Seminar /Conference, where International Consultants/World Bank/International Oil companies and representatives of our own Oil companies can be invited to deliberate and make recommendations.

Foreign Basins

Some of the major Industrial countries who have limited Hydrocarbon resource, have encouraged, supported financially and politically, their oil companies, to acquire acreage abroad to meet the Oil security needs. TOTAL of France and ENI (AGIP) of Italy have major oil interests in the world, and have covered oil supply security interests of their countries.

Oil & Natural Gas Corporation is the major national oil company of India and contributes to over 90% of current domestic oil production. It possesses experience and vast reservoir of technically competent experts to undertake exploration and production activities. The wholly owned subsidiary of this company, ONGC Videsh is already operating in Vietnam, has been entrusted with the responsibility of globalising E & P activities of ONGC.

Since an organisation for global E & P activities is already existing and has a track record of success, in this sector, it would be appropriate to identify ONGC Videsh as the main vehicle for globalising E&P activities for the country.

70% of the world oil reserves are located in Middle East, former Soviet Union and Africa. Exploration and production activities in these areas will serve our oil supply security interests.

In view of the surplus oil situation and low prices, the International oil companies are reducing the investments in exploration. **The time is opportune to make a major thrust, through ONGC Videsh, to acquire more exploration and production interests abroad.** The activities abroad should be supported by the Government and all the Oil companies both in the Public and Private Sectors. A suitable mechanism can be worked out to provide ONGC Videsh with venture capital and manpower support to operate in the competitive environment abroad.

Source: 'Strategy for Oil Security', 23 November 1998

I have relentlessly pursued in convincing others that restructure and integration of the oil sector was the pressing need of the day. Presented here is the enclosure to a letter addressed to Dr S. Narayan, secretary petroleum.

In line with the technological breakthrough, particularly in the field of Information Technology, the Oil Companies in the western world are consolidating restructuring and Re-engineering to reduce costs. Mergers of British Petroleum and Amoco is the case in point. The thrust is to have large Integrated Companies (not in terms of manpower), upstream and downstream. ...

What I always have stressed is the need for the oil sector to be self-reliant particularly more so in the context of the impact of globalisation. The extract below reiterates this point further.

Globalisation Impact

As a result of globalisation and optimum utilisation of information technology, mergers and acquisitions became necessary for survival and growth, as margins had reduced. British Petroleum acquired Brit Oil in 1987, AMOCO in 1977, ARCO (Atlantic Richfield) and Burman Castrol in 2000. It is the largest company in Britain and the third largest in the world. EXXON merged with Mobil in 1999 and it is the largest company in the world. TOTAL merged with Fina and Elf in 2000. CONCO and Philips merged in 2001. These are giants with strong, technological and financial muscles and have assets and operations in many countries.

To stand competition from international companies, Indian oil sector merits analysis and review with a view to reorganise. Piecemeal ad hoc actions to privatise the public sector enterprises whether in refining, petrochemicals, consultancy, exploration and production will be detrimental to the oil security of the country. What is needed, is an efficient sector (public/private or joint) to ensure self-reliance in oil, particularly at the times of national crisis. There are a variety of models available, in different countries at different stages of economic development. A suitable model can be adapted to our needs. This is the vital issue for oil security and accelerated economic development to meet the crying needs of the teeming millions of our people and brooks no delay in working out a proper strategy.

Source: 'Oil Security Walk-up Call', *Kaleidoscope*, 2004.

To further explain the significance of a consolidated oil sector, I had also written to the minister of petroleum. Reproduced below is the letter addressed to him.

20 September, 2004

I had watched and listened with great admiration, the presentation of your vision to meet the socio-economic objectives through the restructuring of the Public Sector Oil Companies. ...

Integration will result in concentration and acceleration of exploratory efforts in 20 virgin basins in the country and basins abroad. You had very rightly emphasized the role played by the Public Sector Oil Companies (OPSOCs) in making six sedimentary basins highly prospective. Exploration risk in these basins is minimal and needs to be exploited fully by the PSOCs. However some of the discovered fields were literally gifted away to the private sector. The Late Shri Rajiv Gandhi (former Prime Minister) had acclaimed the efforts of the ONGC through a letter dated 24 Feb, 1989.

May I reiterate some points that I had made during the discussions on September 1st, 2004. Multinational oil companies aligned with their national interests, have consolidated their positions through mergers and have acquired very strong financial and technological advantage through economies of scale and are diversifying globally to gain first mover advantage through well thought out energy strategies. For the very survival of Indian PSEs in the national and international arena, their integration brooks no delay. This is vital for oil & energy security and economic development.

The eventual form of the restructured entity needs to be discussed, brainstormed and finalized. There are many successful models to choose from to adapt to our needs. Possibly the Supervisory Board model with a number of subsidiary operational companies in various functions (exploration, production drilling, refining, technical services, R&D and emerging technologies, petrochemicals, power generation, gas, transportation, marketing, etc.) could be developed.

This vision will not be without its challenges, as the proposed integration of the Public Sector Oil Companies will pose a major challenge in the Management of Change. People, particularly the senior functionaries will get unduly threatened, for the fear of the unknown. This may in the short-term create an environment of uncertainty. This can no doubt be managed through a proper strategy, close communications and a well developed action implementation plan, all of which can be worked out through the formation of a mandated Advisory Council.

You have very rightly initiated a dialogue through a limited forum to generate discussions to get feedback for implementation of your vision. A number of workshops and conferences may have to be organized at various levels to bring home the challenges and options.

On retrospection, it is gratifying to note that many of the initiatives undertaken and suggested more than two decades back, have now started bearing fruit. At the same time, a man ahead of his time and who tells the truth first has to be necessarily hanged as per a French song, or at least be attempted to be hanged. I was no exception to this rule where such denigrations were attempted, albeit in a disjointed fashion. Besides the attempted hanging, if I may say so, the most painful episode had been, the gifting away of some of the discovered oil fields of ONGC to private players with the favoured stipulation of international price for oil they produce from these fields, while, ONGC was denied even a reasonable price (production cost plus a certain percentage of return) for the oil it was producing. After all, ONGC could only cry at the shoulders of an unresponsive father. This gifting away, however, happened two years after my departure.

19

Oil—Economic and Political Weapon

Background

It is to say the obvious that oil after the formation of OPEC in 1960 and the Oil Price Shock of 1972 has become a political and economic weapon. Some thinkers call it WMD - proved by the conflict ridden history of Middle East and recent destruction in Iraq caused by the power hungry USA, whose appetite for oil is the highest in the world, and is feverishly working to bring maximum oil resources of the world under its influence.

Crude oil prices are hovering around $ 50 a barrel, against around $ 10 a barrel in 1988. Some agencies have predicted that the prices may touch $ 80 a barrel in the near future. It is a warning that days of ready availability of cheap oil are over. The high volatility in crude oil prices is a matter of great concern, particularly for developing economies like ours. It is a wake up call for our policy makers as our planning is based on easy availability of cheap oil, in spite of many oil price shocks in the past and warnings by knowledgeable people.

World Oil Reserves

On the basis of the data published in "BP Statistical Review of World Energy" world's oil reserves should last for a little over 40 years based on the present rate of consumption. This does not take into account the increase in demand in the Third World developing economies. There has been a decline in exploration efforts. The success ratio in finding oil has also come down. The access to new reserves is also being restructured due to rise of resource nationalism by some countries. This has resulted in cash rich oil companies looking for acquisition of small oil companies in the stock market rather than spending efforts and resources in finding more difficult and expensive oil.

A few extracts from Tel Trainer's article 'The Death of the Oil Economy' (*Earth Island Journal*) are as follows:

"The world consumes 23 Billion barrels a year, but the Oil Industry finds only 7 billion barrels a year.

Beyond 2005, the energy required to find and extract a barrel of oil will exceed the energy contained in the barrel"

The picture which has been projected is dismal. The reality may be better. The OPEC members are sitting on huge reserves of unexplored oil. A lot more oil is still to be discovered in many other countries such as Russia and former members of erstwhile Soviet Union.

Oil Prices

The future of oil price is as unpredictable as the results of oil exploration and estimation of oil reserves which are always speculative. The price volatility is due to political uncertainty in the Middle East, increase in demand in developing economies, continuing terrorist attacks in Iraq and Saudi Arabia. Events such as the crackdown on the Russian Oil Company Yukos, and civil strife in Venezuela and Nigeria create the fear of disruption in supplies. Some pundits, as reported in *The Economist*, think that the fear premium may have added US $ 7 to $ 15 to the cost of oil on future markets. Some believe that high oil prices may be caused by the activity in the Stock market and influence of hedge funds. The speculative bubble could burst quite suddenly.

Oil Demand Management by the Developed World

The developed world, after the oil price shock of 1970s, had introduced policies to ensure energy efficiency. USA regulated its automobile industry through the Corporate Average Fuel Economy Law (CAFE); Japan and Europe also reduced energy demand through taxes. There is a constant drive to bring in more energy efficiency measures and to replace oil with other energy sources. Another major factor responsible for reduction of oil intensity in OECD countries in their shift from manufacturing to services.

OPEC Fears

Reduction in oil demand creates its own fears on OPEC members, particularly in Saudi Arabia. One of the oil ministers of Saudi Arabia had warned OPEC members saying, "Stone age did not come to an end due to shortage of stones." Saudi Arabia has always been the voice of moderation within OPEC, as they believe that OPEC members thrive on the economic growth of others, which is concomitant with energy demand. It is in their own interest to maintain control on prices as their revenues depend on stable oil demand. They need revenue to provide social security for their people to which they have got accustomed, through the benefit of black gold bestowed on them by nature. In spite of this, oil prices will continue to be volatile and unpredictable.

Growing Energy and Oil Demand in India

India is on the threshold of accelerated economic development in every field of activity to meet the needs of millions of our people living below the poverty line. India also has the world's fastest growing automobile market which is further driving oil consumption.

Oil is being used as a swing fuel to compensate for inefficiency of other energy sources. In view of this, the planners have got used to planning on easy availability of oil, disregarding the economic realities and world oil scenario. India Hydrocarbon Vision – 2005 document does not have a word about oil price volatility and the need for demand management. The estimated crude requirement of 364 MMT projected for the year 2024-25 is totally unrealistic and overlooks possibilities for conservation and use of alternative fuels. India is now importing more than 70% of its demand. The indigenous oil production has been stagnating since the early nineties. This calls for review of the strategies to ensure oil security.

Accelerate Exploration within the Country

The need to accelerate exploration, to convert balance over 22 billion prognosticated hydrocarbon resource base, into Geological in-place reserves to meet the growing demand for high cost oil cannot be overemphasised.

In spite of financial and technological muscle of a number of international companies who had operated in India since 1867, the year of the first oil strike in Assam till Independence in 1947, oil had been discovered only in Assam. It was a considered opinion of their experts that India did not have oil in any other part.

> Oil is a sensitive and precious economic and developing resource.

By aping the Western model, we have already compromised oil security by gifting away discovered oil fields of NOCs, in highly prospective basins, made so by the sweat and blood of the employees of NOCs. Not one discovery has been made in the virgin basins since the involvement of private sector. Two discoveries, one of oil and other of gas have been made in areas which had been made prospective by the exploratory efforts of NOCs. This strategy of robbing Peter and paying Paul has demoralised the Public sector employees. Oil is a sensitive and precious economic and developing resource. It can't be handled and traded like other commodities.

To further emulate on this issue an extract from the 'Foreword', I had written for ONGC's energy review is reproduced here.

> India has made rapid progress in the economic sphere since independence and energy sector has played a significant role towards India's economic growth. The planning process in India has given due emphasis to the energy sector. 7th Plan envisages 30.48 per cent share for energy sector as compared to 10 percent during the 2nd plan. In spite of these efforts we have still to go a long way in meeting our minimum energy needs. Therefore to make available the desired level of energy input at acceptable costs assumes national priority.
>
> Over 90 per cent of the world's energy demand is currently more than one-third of the energy needs by the turn of the century. However, the task of replacing oil resources is likely to become increasingly difficult and expensive. The prime national objective, therefore, will have to be 'more away from dependence on oil'.
>
> Our energy requirements will have to be viewed in the long-term perspective. We will have to take a close look at the prospects of various energy options, both renewable and non-renewable, to arrive at an optimal mix. A long-term 'integrated energy plan' is required to be drawn specifying the inputs required in each sub-sector of energy.

85 per cent of the oil resources are consumed in the transport and domestic sectors. Therefore, any effort towards reducing our dependence on imported oil will have to be directed towards finding ways and means of substituting the use of oil in these sectors by alternate sources of energy.

Natural gas with a vast resource potential holds a good promise as an alternative to oil. Natural gas utilisation is on the increase worldwide, specially for fertilizer, power and petrochemicals industry. In India, natural gas reserves have increased several folds over the years and are expected to increase yet further. The role of natural gas in the future energy balance is expected to be significant.

The energy industry today is facing new challenges. This is the most opportune time to reassess policies which have guided our developments in the past, to review contributions made by new technological developments and to accordingly formulate our future strategies. A multi dimensional approach is required to meet this challenge. Firstly, a long term Strategic Plan should be drawn for building up petroleum reserves both via domestic exploration as well through exploration ventures abroad. Secondly, a time-bound plan for substitution of oil with alternative sources of energy and increasing the share of non-conventional energy forms should be formulated. Thirdly, we should have a deliberate and implementable plan for conservation and efficient use of energy. Effective 'demand-management' is the 'key word'.

I hope this perspective devoted to 'Future Energy Options' would provide energy planners with new insights into the nature and magnitude of various issues involved and help them evolve strategies to be followed to meet our growing energy needs.

Source: 'Future Energy Options' December 1987 – ONGC Environment Scanning Group Report

This sets the pace for future initiatives in the energy sector. The ONGC Environment Scanning Group had produced many reports. The contents and recommendations of these reports are valid even today.

Foreign Basins

ONGC through energetic efforts has made inroads into the domain of multinationals and has acquired equity in a number of basins abroad. The progress of these projects needs to be followed up with great seriousness. There is need to learn from the sad experience of the Vietnam concession obtained with lot of efforts in the late 80s. **The major interest had literally been gifted away to a multinational in the early 90s.**

Recent Initiatives

The Oil Minister is personally directing initiatives to build relations with oil/gas producing countries for long-term oil/gas contracts. Major pipeline projects for import of gas from Iran through Pakistan and from Myanmar/Bangladesh are being discussed. These, no doubt, have associated serious security risks of crippling the industries using this gas in the event of any disruptions due to militancy or other political reasons. To use imported gas, national gas grid will have to be established. Huge gas reserves in Tripura also need to be taken into account for the gas grid. The conceptual National Gas Grid plan which was prepared in mid eighties by ONGC needs to be updated.

Energy Diplomacy

The energy diplomacy initiated through Asian Round Table Conference on oil cooperation is bound to create good relations for getting favoured treatment at some stage. These initiatives have potentialities for results but may remain only dreams if not followed up with vigour to handle the management complexities involved. It also calls for unstinted cooperation within the various wings of the Government. **In fact, we have the knack and history of converting simple into complex issues through bureaucratic and political hurdles.**

The initiatives taken by the Minister of Petroleum for supplies of gas and crude through pipe lines from abroad are long-term measures for energy security if at all they materialise as they are pregnant with many political considerations and military threats. The long-term contracts for LNG will surely give a lot of confidence for oil security.

Re-engineering of Oil Sector

The initiative by the Government to set up a committee 'Synergy for Energy' to reengineer the fragmented oil sector is commendable. There are several successful organisational models available from democratic countries having a mixed economy environment. A matrix organizational structure through the formation of a Supervisory Board/Holding Company, which a number of operating companies with full administrative and financial powers will ensure the desired results for different oil businesses.

Oil Security Priority

Oil demand management should be the highest priority through conservation and efficient use. An integrated energy plan with an objective to move away from dependence on oil needs to be prepared and implemented apart from accelerated exploration and other issues detailed earlier to ensure rapid economic development.

(Based on article with the same heading published in *Economic Journalist,* April-June, 2005)

20

Vision for Tomorrow

We have been in an era of rapid change witnessing radical political, economic and technological changes, all over the world. The environment of liberalisation and globalisation have largely influenced the working of Indian business enterprises. To ensure the survival and growth of enterprises and to make necessary changes to business plans, it is essential to continuously scan the changes occurring in business environment within the country and abroad.

As a responsible corporate entity, therefore, ONGC took on itself the responsibility of projecting to the government the issues and plans meriting attention for energy security. No doubt these would also have bearing on the oil sector in general and ONGC in particular. To meet this objective ONGC had to mobilise the combined wisdom of rank and file and scan the external environment and plans of international enterprises in the field of oil and gas.

Creating Awareness and Thinking Ahead

The 'Economic Environment Scanning Group' of ONGC produced publications on the energy sector such as *Energy in Profile, Future Energy Options, Gas,* etc. **Actions were taken within the limited powers of ONGC, and the challenges for the future were projected in the ONGC annual reports.** Other issues were projected through various channels, forums and documents to generate discussions among the public and government circles as would be seen in the following pages. Looking back it is seen that our plans and thoughts were much ahead of times as actions by the government took place slowly, years after our projections, and some are still under consideration.

Views of a Former Member (Personnel)

Dr S. Ramanathan (former member HR and an eminent geologist) had stated as follows on 18 April 2003 in ONGC conclave at Bangalore, almost 13 years after my retirement.

> With the embarrassment of riches of Bombay High firmly established, the organization graduated to adulthood. It had the benefit of a benevolent M (P) and a chairman, who was Chairman (HR), par excellence, in addition to his other functions as the Chief Executive. It will be difficult to enlist all the changes made during this period but I would mention a few all the same.
>
> - The vision and mission of the organization were clearly defined for the first time and became a household slogan of every employee.
>
> - Within three years, the breaking of every rule and regulation of the R&P —Recruitment and Promotion (1980) was a rule rather than exception, and for right reasons, keeping in view the need of the organization for rapid development and to make the best use of the in-house personnel.
>
> - When I occupied the chair of M (P) in 1986, I was thanking my stars that I was not hauled up for contempt of court. When the author of R&P (1980) visited ONGC in 1988 and found the changes, he admired the development and conceded that the **only permanent thing is CHANGE.**
>
> - **Col. Wahi was the precursor of privatisation of non-core functions in ONGC.**
>
> - Indigenisation, including supply boats with Indian flags and Indian owned contract rigs got a massive thrust. Today, these companies are earning foreign exchange for the country by operating overseas. Not only well platforms but also production platforms and water injection platforms were fabricated indigenously.
>
> - Business Group concept and cost centres were introduced. *But, he was ahead of his time* and, therefore, it required a Mc.Kenzie to push it forward with other adaptations of the day.
>
> - *He was a firm believer that an organization or an individual, which does not grow, stagnates and finally decays. And, anyone who ceases to improve, ceases to be good.* The organization's growth curve and the individual's growth path were always seen together and were on the ascendancy.

The Bombay centre, which was labelled as a superior offshore culture was implanted onshore by transfers and vice versa for cross culture fertilisation. **This benefited all by the Common Basin Approach.** The geoscientists from on-land could exploit the advantages of a common basin approach offshore, and the drillers, for instance, could so easily transfer the inclined well drilling technology to ER (Easter Region) for development wells that made a paradigm shift in drilling speed and cycle speed. This helped achieve higher production, without having to address to environmental problems.

It is generally believed and accepted in the country that Western India is a decade ahead of Central India, which is again ahead by a decade of eastern India. The glamour of presentation of data, inculcated from CFP, (A subsidiary of Total of France) which was till then proprietary to BOP, (Bombay offshore project) was transferred to other regions. Yet, the first power point presentation in the commission was given, not in Bombay nor in the HQ, but in ER, albeit piloted by an ex-BOP executive, Mr Atul Chandra.

- Long-term planning was done for the first time where, in the organization, the individuals could see their own future intertwined.

- **He (Col. Wahi) taught us how to convert even a threat to an opportunity.** I would only recall the Sagar Vikas blow-out. The first major blow-out of that magnitude offshore was questioning the safety standards of ONGC. The way in which press was handled by regular daily briefing and presentation, has remained a model for PR handling and quoted as a case history in mass communication. I dare say this has not been surpassed even by CNN in the handling of the Iraq war, since in the latter case, the credibility has been very suspect. The South Kadi blow-out onshore was also handled equally adeptly, though by adapting different techniques.

- Transparency and information sharing helped easy approval of projects like BHS (Bombay High South) development, introduction of horizontal drilling technology and conversion of Bassein Development platform proposal from sweet to sour system on the basis of new evidences, through the PIB (Public Investment Board), which was otherwise initially very hostile to these proposals. I had the good fortune of being associated with these projects, in getting the approvals.

Vital Issues

Issues which were outside the control of ONGC but were essential to revitalise energy security and were projected included: an integrated energy plan; restructuring of the oil sector particularly public sector enterprises; growth of energy sources especially gas; development of alternative sources of energy; purchase of high technology companies from developed countries; exploration in foreign basins; National Gas Grid; and creation of a subsidy fund to reduce financial burden on the PSEs and promote indigenous development.

Some of the main initiatives which ONGC had taken to be the role model in efficient operations and were appreciated by media and government, and got reflected in ONGC's annual reports include: long-range planning; use of information technology; setting of clear mission and objectives; outsourcing of non-core activities; promoting new concepts in oil exploration; **introducing a collaborative culture which ensured total commitment of employees** and outside agencies including state governments; laying thrust on research and development; emphasising training and development particularly leadership development within the organisation; meeting corporate societal responsibilities of the unprivileged groups and widening the infrastructure by developing the sites of operation. In the following pages these unique changes initiated by ONGC visualising its phenomenal growth have been presented.

Paradigm Shift

Several policy changes had been implemented in the 1980s. The extract below throws light on the heights attained by ONGC as the harbinger of growth and improvement.

The Eighties Belong Substantially to ONGC

The Eighties belonged substantially to the Oil and Natural Gas Commission. Over these years there had been a three fold increase in production. During this decade, these have their impact on reducing the trade gap, thus providing a strong support to the opening up of the economy. In 1986, ONGC was ranked the ELEVENTH among the profit leaders in the corporate world outside us.

Col Wahi has been an outspoken critic of the way the bureaucracy interfered with the public sector units. In an address at the MMA annual management convention held in last December, Col Wahi said, "The country has chosen the path of mixed economy for development, where government machinery has a direct control over planning, investment priorities, clearance for various activities, inputs and policies, which have direct impact on time and cost, performance and productivity." Business and government have to be partners in the economic development of the country to achieve growth with the optimum level of productivity. Do we have such an environment in the country? Is the government organised to play this role? The answer is 'No'. The bureaucratic delays have a snowballing effect all round, resulting in low productivity. **The whole structure of management of business, particularly the public sector, merits a detailed review and major structural changes are required to improve productivity.**

Inefficiency is very infectious. In the existing system, there is need for a statutory audit to identify bureaucratic delays and inefficiencies. I have no doubt that business would be too happy to finance such audits!

A public sector chief executive can be dropped if he did not perform; but no secretary to government can be sacked for non-performance; at best he can be transferred back to the state or to some other department. Unfortunately most bureaucrats are not goal-oriented. The power of the economic ministries without accountability is at the root of delays and indecisions.

Source: S. Viswanthan, *Industrial Economist*, April 1988

Under my chairmanship ONGC was instilled with the courage to dream and aspire new heights. Some of the views mentioned in the letter reproduced here are now being expressed by the higher echelons of the government to allow mergers and acquisitions within the public sector enterprises and operate in the capital market.

Organisational Structure and Linkage with Ministry

There are more than 209 public sector enterprises, operating almost in every sector of the economy. In the same sector of the economy, more than one PSES are operating. Their working is being coordinated by the respective administrative ministries, rather ineffectively due to different objectives of the administrators and public sector managers. The role of the administrative ministry is to 'administer' whereas the role of the corporate enterprise is to 'manage'. Public sector has to work on commercial lines which involves speed in decision-making, calculated risk taking and innovative approach for achieving quantifiable results....

The public sector enterprises in the same sector should be controlled by a holding company and the Chairman of the holding company should report to the minister directly and should also be designated as secretary to the government....

The concept of the holding company will ensure vertical as well as horizontal integration of operations in each sector of economy, resulting in optimisation of resources....

Diversification: Public Sector Enterprises should have the flexibility to diversify on the lines of ENI Italy. Example of a large number of successful companies could be cited who have adopted this approach.

Performance Appraisal: The performance appraisal of Public Enterprises should be on profit (and) based on international norms. The financial disadvantage to the company as a result of government directive for meeting certain political or socio-economic obligations should be suitably provided for in the Annual Accounts of the company.

Source: Do No. 4/62/84/CH dated 19 December 1984 addressed to Dr Sengupta, special secretary to prime minister.

*****

Efficient Demand Management Policy Vital: Wahi for Integrated Energy Plan

The growing energy needs of the country could be successfully met through an efficient demand management policy, which remained unattempted in any methodical way for well over four decades according to Col. S.P. Wahi, chairman, Oil and Natural Gas Commission (ONGC), reports PTI.

In an interview with PTI, the ONGC chief underlined the need to draw an integrated energy plan keeping in view the possibility of 65 million tonnes of oil production as projected in ONGC's corporate plan and meeting the balance requirements through other forms of energy.

Countering the caution that self-reliance in oil would go down as 'prematured', **Col Wahi contended that it was wrong to look at oil 'in isolation'. As such, proper energy mix should be worked out if the energy scenario in the coming years were not to turn dismal**, he said.

Noting that ONGC does not manufacture oil but locates it, Col. Wahi asserted that the success ratio between the number of wells in which oil and gas was found against the number of wells drilled is 1: 3 as compared to the world average of 1: 5.

He said during the current plan the stress has been shifted from production to exploration so that addition to reserves could be registered. While there has hardly been any proven method of direct detection of hydrocarbons other than through drilling, Col. Wahi deplored that drilling density remained dismal in the country.

The drilling density is around 11.6 wells in 10,000 sq. kms, which is very 'low' as compared to the global average of 100 and around 450 in the U.S., he said.

He said the immediate and medium term prospects of ONGC's operations were 'bright'. Besides, in tune with the ONGC's practice of reinforcing 'success', crash programmes were drawn for promising discoveries for early development of field and for generating quick returns, Col. Wahi added.

When asked about the 'scarcity price' for oil being charged by the government from the consumers to contain the consumption petroleum and petroleum products, Col. Wahi conceded that the benefit of this accruing to the exchequer has not been passed on to oil companies like ONGC.

He further noted that the administered pricing mechanism was introduced by the government in 1980-81 when the international price was $34 a barrel. ONGC was paid a net price of $10 (Rs 976 per tonne).

Even today the same price of $10 is applicable although the Gross Domestic Price has increased from Rs 1,182 per tone in 1981 to Rs 1,831 today. He said the increase was on account of statutory charges and 'ONGC does not get any advantage'.

Col. Wahi said despite the stagnancy in the administered pricing mechanism for the past several years, the internal resource generation in ONGC has been progressively rising and is about 90 per cent now.

Col. Wahi suggested the creation of a national subsidy fund to encourage indigenisation to achieve self-reliance in oil related equipment, materials and services.

Pointing out that the ONGC is a commercial enterprise and is subsidising the indigenous oil field equipment industry in a big way,

Col. Wahi said that the Commission had spent Rs 26 crore in this direction in the last three years.

This, he said, added to ONGC's cost of operation with concomitant, curtailment in profits. Hence, it is 'essential' that the government considers setting up a separate subsidy fund for promoting indigenisation effort, he added.

Col. Wahi said Gandhar field, located about 80 kms south of Baroda and 37 kms, north west of Baruch, is fast 'emerging as the most promising onland field'.

With the resource base of 466 million tonnes, Gandhar field is likely to pan out about four to five million tonnes of oil per annum, he said.

Source: Economic Times, 17 August 1988

Long-Range Planning

A culture of long-range planning was created within the ONGC. Inherent in the new exploration policy was the urge to look beyond the Annual and Five Year Plans **to ensure that long-range objectives were not sacrificed for short-term gains**. The Commission broke new grounds by formulating a long-term operative plan for 10 years based on a 20-year conceptual plan. ONGC's Annual Report 1988–89 stated, "The long-term Corporate Plan is being updated on a five-yearly basis so that the Plan for 20 years hence, is available. A Plan covering the period up to 2014–15 is under finalisation." The following extract further throws light on the initiative taken by ONGC.

> Colonel S.P. Wahi set up a task force to draw up a 20-year oil exploration and exploitation perspective plan aimed at chalking out a blueprint strategy for national self-sufficiency in crude oil production. An initial draft based on Seventh Plan projections was revised in 1985 even as the plan was being finalised, and a new document—Revised 20 Year Perspective Plan (1985–2005)—was drawn up in December 1984. Although this revised plan is still not the final word—and neither can it be in the game of probabilities which is oil exploration—it has nevertheless set in perspective both the vision and the objectives of ONGC....

Source: Sujoy Gupta, 'Wahi's Ambitious Perspective Plan', *Business World*, 18–31 August 1986.

Changes in the Organisational Structure

Keeping in view the long-term objectives, future growth and challenges, as mentioned in earlier chapters the organisation was restructured on '...a common basin approach and the concept of Business Groups with centralised policy making and decentralised administration, with commercial relationships among these groups' (Annual Reports ONGC). This concept yielded excellent results. The need was felt to review continuously the organisational structure and adapt it to the changing needs and situations. The organisation was to split into subsidiary companies to further optimise the resources. As a first step in this direction, a number of profit centres were created. **It was decided that the research and development institutes would also act as profit centres** and would mainly focus on meeting the current needs of the operations.

Emphasis was also laid on the need of interaction with universities, IITs and other scientific organisations to meet the basic research needs, so that R&D institutes of ONGC could apply themselves to applied research needs and upgradation of technology. The integration of computers and communication systems was given a bigger thrust for the collection, collation and dissemination of information.

New Thrust Areas

Historically ONGC was having under its control a number of low technology areas and activities which would best be handled by other organisations within the public or private sector, so that ONGC's management time could totally be devoted to high technology areas related to oil exploration and exploitation. With a view to ensure that growth in support services did not hinder the progress of exploitation and drilling operations, a policy decision was taken to shift the growth and development of support services as also other low technology areas to outside indigenous parties. The formation of cooperatives by the employees of the Commission, by taking premature retirement, to provide services to the Commission was encouraged.

Overseas Operations

In ONGC, a few had the vision to comprehend the phenomenal economic and political potential that the oil sector in India held for itself. A few at ONGC felt the need to acquire high tech companies. I had commented at one instance, '... **We should also scan the international market for acquiring high technology companies, as this is an opportune moment to make such investments**' (*Source*: ONGC Annual Report 1985–86). However, due to bureaucratic hurdles such opportunities were lost.

Worker's Participation in Management

A unique experiment in the history of the public sector undertakings in our country has been the formation of advisory councils in the areas of exploration, human resources development and management. The advisory council on management and corporate policies was aimed towards evolving the right view of management policies—**a novel attempt by ONGC towards labour participation in management**. To promote transparency, two members from the media (*Times of India* and *Business India*) were also made members of the advisory council, apart from the presidents of all the unions.

Exploration and Production—Energy Security

ONGC had made remarkable progress in oil exploration. The discovery of oil at Bombay High gave a new dimension to the exploratory effort. The prognosticated resource base was earlier assessed being limited to water depths of 200 metres. However, exploration activities soon revealed that deeper waters offered tremendous potential.

The Southern Region had shown very high prospects of hydrocarbons as a result of discoveries in Narimanam, Kovilkalappal, Bhuvanagiri and GS-16 (Offshore). It was therefore, decided to have a flexible plan which could enable reallocation of resources to areas showing more promise. It was proposed to multiply the activity level in Krishna-Godavari and Cauvery basins.

The ONGC Annual Report 1987–88 also emphasised that 'exploration in foreign basins would be extended further as corporate strategy to have control over more oil'. It is unfortunate however, that such hard-earned fruits of ONGC's efforts, at present, have been virtually gifted away. For future prospective it was decided (as recorded in the Annual Report 1988–89) to intensify focussed exploration campaigns abroad, specially in the Third World countries and to develop structured cooperation with SAARC countries in the petroleum exploration, production and transportation sectors. Following up with this strategy, the marketing of ONGC technology was undertaken by the Overseas Operations Group within ONGC. The consultancy agreements with ADNOC (UAE), PETRONAS (Malaysia) and PTTEP (Thailand) brought international recognition to ONGC's scientists and engineers. The first phase of seismic data acquisition in Vietnam was also completed successfully.

It is a matter of pride that all the plans, projections made and actions initiated by ONGC during the period 1981/89 have borne fruits. The Southern Region is one of the most prospective hydrocarbon province and that deep water exploration has also borne results, and major acquisitions for exploration in foreign basins have been made for which Vietnam acreage was obtained in 1988/89.

It is heartening to note that the government has made positive moves to make this idea of collaborative approach with SAARC countries.

Social Responsibility and Environment Protection

The objective of ONGC's community welfare programmes had been the uplift of the socially and economically weaker sections. In addition to adoption of villages and provision of amenities in townships, scholarships for students from Scheduled Castes and Scheduled Tribes studying in engineering and other technical institutes were also instituted. **Above all, environment protection was made the fifth corporate objective. An Institute of Petroleum Safety and Environment Management was also proposed to be set up at Goa** for which land was acquired and a nucleus created. Appendix U.

Special Focus – Uniqueness of ONGC

ONGC mode of operation has always been different from other public sector organisations. Its foremost strength was that 'far from being a ponderous bureaucracy', ONGC was 'a fast paced technology oriented company' (Sujoy Gupta, *Business World*, 18–31 August 1986). "While self-sufficiency in oil by 2005 is our number one objective, close on its heels is objective number two: self-reliance in technology," observed member (exploration) S.N.

Talukdar, a competent geologist associated with ONGC since its incorporation in 1956.

Another strength which differentiated ONGC from the run-of-the-mill 'babu' dominated government company was its fast decision-making processes. Hence, emphasis was laid on an efficient system of communications. Dedicated hotlines covering 23,100 km coupled with 1,502 very high frequency (VHF) wireless sets connected the headquarters, regional offices and drill sites six days a week. In addition, an elaborate satellite communication system was also installed to guarantee instant contact between the headquarters in Dehradun and even the most far-flung rig on land or sea.

Computerisation of operations was prioritised. An entirely computer-based management information system (MIS) was introduced, and project planning, project appraisal and control were computerised.

Adding to such innovative methods were the task force concept, according to which a specialised task force of experts were assigned a deadline to go into the details of any specific problem in addition to their normal work, and come up with alternative solutions (out of the box) for implementation. Their recommendations were then quickly vetted by the executive committee consisting of the chairman and the full-time members of the commission and instructions were issued. The concept paid rich dividends in improving efficiency.

The other innovative concept was that of 'management by wandering'. **Accordingly those responsible for decision-making were encouraged to move out of their offices and interact with the men on sites.** Senior managers were encouraged to tour to the forward areas and hold meetings right there on the spot, instead of asking people to come over to the regional or head offices. Helicopters were provided to the regions for fast movements of the senior executives to give support to the operations.

New ideas and initiatives such as national gas grid and Five-Year Plan on environmental management had taken shape. Instead of the initiative in the hydrocarbons sector being with the government, ONGC advised the government on what ought to be done.

Source: Sujoy Gupta, *Business World,* 18-31 August 1986

Collaborative Approach with Government

ONGC possesses the required technological ability, management sensitivity, perception, innovativeness, and above all, willingness and commitment to manage the changes. The key players of the oil game—the ONGC, the government and the administering agencies are required to effectively play their desired roles in complete harmony, and with an obsessed sense of total commitment to achieve their respective goals on time. Such a vision for tomorrow can only be nurtured under the wisdom of true leadership at all levels in the business enterprise. Therefore, development of leadership is important for success during good and not-so-good business environments. (ONGC annual report, 1988-89).

Leadership Development

What is Leadership?

Leadership is an ability to inspire a group of people to move willingly and enthusiastically towards a common group objective in a synergetic manner. To inspire, the leader has to be a role model. For this the leader has to have character, competence, concern for the organisation and the people, commitment to achieve results, courage to take decisions and lead from the front, the ability to conceptualise so that woods are not missed for the trees, and the attitude and ability to communicate.

For optimum performance of an enterprise, structure, technology, processes, systems, procedures and other tools of management are important, but most important are the morale and motivation of employees to ensure collaborative approach and sense of belonging to the enterprise and relationship management with the external environment. **Therefore leadership development at all levels is vital for success.**

Leadership Abilities

Leaders have to have the vision to steer the organisations to the path of growth and have to ensure stability during difficult and turbulent economic times and crisis situations. Every economic decision has a risk. Leaders have to have entrepreneurial spirit to take calculated risks in time and to make the future, rather be taken over by the events created by the economic, political and technological changes.

Human relation conflicts are the root cause of the low performance of many enterprises. **The leaders have to have the knowledge of human psychology and have strong emotional intelligence, to identify the conflicts and resolve them.** The leaders have to know their people well, their dreams, their concerns and their strengths. Leaders have to ensure a collaborative culture in the organisation to ensure harmony and a good working environment, so that everyone has a sense of belonging to the organisations and have a Espirit-de-Corps, This enables everyone to operate to their fullest potential with vigour and passion. The extract of an interview as appended below brings out the role of leaders towards their junior colleagues.

Incompetence is Product of Organisation Culture

How do you define incompetence?

Incompetence is generally related to the performance expectations from the managers. There is an urge in every individual to be accepted as competent. The performance however, depends on the professional competence of the individual and the organisational work culture which is created by the senior corporate leadership, particularly the chief executive. ...

So do you feel there is no such thing as incompetence?

During my work experience of over 47 years, I can identify only few managers who could not be motivated to put in their best. The organisations have to ensure continual training and development, introduction of the state-of-the-art technologies, particularly the information technology. The organisations have to be engineered in line with the international development to ensure a healthy pressure on the managers for performance and competence.

What do you have to say of managers who are dubbed as incompetent since they are not able to meet the targets?

Achievement of targets is not entirely dependent on competence of any one manager, particularly when all inputs required to meet the targets are not under the control of that manager. The manager may have the potential to meet the target but the performance may not be achieved due to factors outside his/her control. There can be possibilities of 'square pegs in round holes', when redeployment may be essential to achieve the targets. The organisation's culture and structure have to ensure working through the strengths of the people and team working. ...

Do you think people get labeled as incompetent because they are unable to meet certain unrealistic targets set by their supervisors?

You have to set high targets only then you will get high performance. It is better to get 90 per cent of a high target than get 110 per cent of low target. But merely by setting high targets, an organisation cannot absolve itself of its responsibilities, it has to supply the wherewithal to achieve the targets.... Initially workers might resent high target but over a period, they get used to high performance standards.

Source: The Observer of Business and Politics, 18 June 1997

Leadership Attitudes and Tactics

Leaders have to have the flexibility to change plans to meet the changing situations and the strengths and concerns of the team. They work through the strengths of the people and create a work environment of development, mistake, tolerance, friendliness, fairness and firmness. **All decisions have to take into account the morale and motivation of the employees and money, the bottom line of the organisation.** Leaders should ensure that group and individual performance be recognised and rewarded. A sense of belongingness to the organisation and Espirit-de-Corps needs to be created, and resources optimised to achieve organisational excellence. The leaders have to know their people well; their dreams, concern, development and growth have to be the concern of leaders. A very close link also needs to be maintained with the retired employees and their families.

Executive Mix

The experience of working with a variety of organisations in the public and private sector indicates that most organisations have four types of executives. The majority of them comprising 80 to 85 per cent have a high degree of managerial and professional skills. They have answer for every problem but are afraid to act. Mostly they act on orders from above or pressure from below. They are the backbone of the organisation but only require a motivating working environment. Among the rest, 5 per cent come under the category of specialists who attach a great emotional value to a particular function/technical aspect and strive to attain it to perfection. They are of great help to undertake research in their areas of specialisations and can provide answers to basic problems.

There are others, constituting less than one per cent of the executives, who have only peripheral, organisational and professional knowledge, but a high level of interpersonal skills. This group excels in spreading data, mostly rumours and can be constructive if identified and managed well. The remaining 9 per cent of the executives have leadership abilities and are the driving force behind an organisation's success. They are the leaders, the motivators, who lead from the front, and are the role models for the main work force.

The leaders create a culture of development, which promotes hunger for knowledge, information and latest technology to keep the organisation globally competitive.

Process of Leadership Development

Schools, colleges and institutes of higher learning spend a lot of effort to impart knowledge, develop managerial and functional skills, but place very little emphasis on character and leadership development for human resource management. **Therefore, there is no shortage of managers, but managers with leadership abilities are in short supply.**

Leadership however can be and should be developed through deliberate training and development in schools/colleges and institutes of higher learning. It is not always inherent by birth and also does not depend entirely on proper upbringing. The training methodology followed at military institutions like the Indian Army can be of great help.

Training and development techniques should create a competitive spirit, and ambition to grow with commitment and achieve excellence. Working in teams brings out advantages of working through the strengths of the team members. Group sports, mountaineering, picnics and hikes are of prime importance to develop leadership with collaborative spirit and courage to meet crisis situations. Study of autobiographies/ biographies of great business and military leaders are of a great help, to learn from the experiences of others. Psychometric testing techniques are very useful to counsel the individual managers for self-development. The commitment to the organisation and its people has to be emphasised, to be a successful leader with name and fame.

Stress on emotional intelligence is of absolute importance to develop proper relationships within and outside the organisation. The societal needs within the limits of corporate governance have to be emphasised during training, as also global environmental concerns.

> The success and failure of an organisation is related to the quality of leadership at all levels.

Employees in the organisations come from different strata of society and different parts of the country or the world. It is important for the leader to bring them up to a minimum acceptable standard of attitudes, behaviour, discipline and norms of performance. This will involve structured attitude surveys, interviews and suitable training programmes, to develop adequate confidence and correct attitude. Executives have to be encouraged to take part in debates and in cultural programmes. Inter-regional cultural and sport competitions help to create competitive spirit, innovation and creativity.

In the formative years of one's career, one does look for role models and do accepts willingly the advice of senior colleagues, who have a good track record and have established a name for themselves. The senior management must spend a fair amount of time to discuss and work out strategies for the development of leadership abilities among younger colleagues. In the in-house training establishments only outstanding executives should be positioned as trainers. As a matter of strategy junior colleagues should be made to attend meetings by rotation where their senior colleagues are members. This is very helpful even for succession planning and understanding strategies for decision-making. A leader has to set an example in time management which is so vital for performance.

The success and failure of an organisation is related to the quality of leadership at all levels. Managerial and functional skills are important, though these can be even outsourced, but leadership to inspire cannot be outsourced and has to be developed. Leadership calls for personal sacrifice to lead from the front, and to have concern for the morale and motivation of the people. More than anything else leaders have to have character and courage to act and not miss opportunities.

Appendices

Appendices

Appendix – A

**DO No. 6/26/ CH. Oil & Natural Gas Commission,
Tel Bhavan, Dehradun-2480003 October 20, 1981**

1. Over the last few months I have been interacting with a large cross section of our people to find out the strengths and weaknesses. I wish to share my thoughts with you and would like to be benefited by your reactions and of your colleagues.

2. ONGC has a record of success but nonetheless the tasks ahead are very onerous and pose a unique challenge. We have the singular responsibility of achieving the objective of self-sufficiency in oil for our country. Our greatest asset is the quality of our human resources. I have full confidence that the faith reposed by our country will be fully redeemed.

3. **Clear Objectives :** Each member of ONGC has to work to clear objectives which will have to be systematically crystallised on scientific lines. Obviously, clear cut objectives, responsibility and accountability at various levels has to be laid down. **A concept of responsibility accounting will have to permeate the entire organisation hierarchy.** As a first step, I would suggest that every executive may be asked to identify his objective and accountability as perceived by him. It is my firm conviction that executive's performance can touch new heights if they are given specific responsibility with matching authority, flexibility and delegation. I am committed to pursue strategy of decentralised administration and centralized policy-making. I would appreciate suggestions for additional delegation of powers to improve efficiency and effectiveness

4. A Management Service group is being established at the Commission's Headquarters and also at the Headquarters of the members to take care of the systems and procedures, organisation development, management, information systems, perspective and corporate planning, industrial engineering studies, management audit & follow up of major project.

5. The volume of business has increased over the years and it would increase manifold in the years to come. The organisational structure would therefore need a thorough review

for proper restructure to meet the needs of not only today but also of tomorrow. **A need for marketing/ commercials department and technical services group to take care of maintenance and equipment management is obvious.**

6   **Common Basin Strategy :** A happy blend of expertise between offshore and onshore has to be ensured both by inter-transfers and continuous interaction. **The basins which extend from onshore to offshore will have to be under the same management for optimum utilisation of slender technical resources and ensure unified strategy for exploration and exploitation.**

7   **Consultancy Group :** A consultancy group will have to be established not only to meet our needs but to enable the commission to extend its services and expertise to the international field. There is a tremendous scope. This will also act as a window to the latest developments and innovations in the field.

8   **Organisation Development :** The organisation development has to be kept under continuous review to meet the growing aspirations of the people in line with the growth of the organisation.

9   Tasks in ONGC call for a high level of professionalism, particularly in scientific and technical operations. It should be our endeavour to maintain a very high level of competency in all areas of work. It would be necessary to constitute groups of officers, with proven high professional standards, to monitor the level of professionalism in various disciplines and make suggestions to the top management for constant improvement of professional skills.

10 **Cost Awareness : Cost consciousness will have to be our culture. An economy** group will have to be set up at all appropriate levels. The Finance department will have to play major constructive role to bring in cost awareness. Normative data will have to be developed for effective control over utilisation of resources. A scientific budgetary control system will have to be evolved so that there is an appropriate awareness that results have to be obtained at a given cost and not at any cost.

11 **R&D - Profit Centre :** The three institutes have to be **the temples of knowledge and excellence** so that not only we avoid dependence on imported technology on a large scale, but are able to extend our activities beyond our borders to derive economic advantage and establish a close and direct contact with the technology abroad as equals. **These Institutes will have to be 'profit centres'.** The work done by these Institutes will have to bear close match with the needs of our various regions/projects. Besides giving economic and commercial advantage, this approach will totally integrate the research activity in the Institutes **with the field needs. Of course, basic research will also be given its due place in keeping with our long term needs.**

12 **Bench Marking :** The optimization of material and human resources must receive the highest priority. We have to ensure the highest standards of work performance - both qualitatively and quantitatively. **This can only be ensured if data of best performance in the world is continually obtained and updated.**

13 **Modernising of Equipment : The updating and modernising of equipment** has to be under constant focus all the time to achieve the best performance standards. The maintenance has to be organised on most scientific lines to get the best out of the equipment. Detailed check lists for carrying out preventive maintenance has to be manualised. The store management is another area which needs immediate attention to avoid build up of inventories. **A task force will have to be formed to arrange the disposal of surplus equipment and stores.**

14 **Equipment Management :** There is therefore, an immediate need to have a look at the existing system of updating, modernising and overhauling of the equipment keeping in view the need to achieve the best results in the light of the developments in other countries. We have to establish a very strong technology development group to look into the perennial problems of the Projects both in terms of equipment efficiency and logistics. The technology development group has to play an effective role in introducing innovation to the existing equipment so as to materially enhance their productivity. Besides, the group must actively liaise with technological

developments so **that new acquisitions are sophisticated and of comparable international standards.**

15. Closely akin to the above issue is our project organisational strength in bringing a proper match between the receipt of new equipment and its utilisation. We cannot afford to waste our financial resources by keeping the equipment idle. All concerned must take energetic measures to reduce the time lag between the receipt and effective utilisation of the equipment.

16. **Technology Transfer :** The contracts have to be monitored continuously to get the best results particularly in the area of technology transfer. A review of existing know-how / technology transfer agreements would be necessary to determine present status. A separate cell may have to be created to provide necessary feed back.

17. **Quality Assurance :** The quality assurance and quality control, particularly in drilling, seismic data acquisition and processing, formation evaluation, cementation, production testing, well stimulation, etc. has to be a part of our system.

    Zero defect philosophy, with its overwhelming emphasis on quality is said to be the secret of Japan's enormous success in a highly competitive world. **The zero defect concept to do everything right the first time has to be our second nature.**

18. **Knowledge Explosion :** We should not under any circumstances resort to short cuts. In a large organisation like ours we cannot be merely guided only by traditions or by total subservience to the rules. The decisions and approach will have to be need-based in a given situation. Suitable system and procedures will have to be evolved and continually kept updated based on our experience and constant exposure to new knowledge/data being generated. There is a knowledge explosion - each one of us has to catch up with this race and remain updated otherwise we would get obsolete. The Commission Headquarters and every Regional Headquarters must have a data bank with proper retrieval system.

19. **Training and Development :** Peter Drucker has described Japanese approach to education very perceptively. He calls it Zen approach, according to which the purpose of training is continuous improvement in performance. It is with this

approach that training needs have to be assessed both in fields of management and technology at all levels. In some areas, we continue to work with outdated methods and techniques. A deliberate effort has to be made to propagate the latest knowledge. Every body can improve his performance throughout his career by continuous training. A culture of self development on a continuous basis has to be imbibed.

20 The project must develop good libraries and ensure training lectures by their own executives to keep their knowledge up-to-date. This programme should be organised on a systematic basis. The executives should be given specific technical subject in which they should undertake in-depth study and then share the knowledge and findings with other colleagues and subordinates. This would tend to keep technical obsolescence at a distance.

21 Action is in hand at the Commission's Headquarters to examine the present status of training at all levels and to work out strategies for the future. **This will continue to receive my constant attention.**

22 **High-tech Manpower Severance Rate :** The problem of shortage of personnel in some critical areas has to be tackled on short and long-term basis – may be by increasing the shift time by giving due compensation to the personnel involved and by adding 10 to 20% more persons in critical areas where severance rate is high. Simultaneously, proper induction plans will have to be developed to take care of wastages on an annual basis.

23 **Exploration Strategy & Deeper Waters :** Petroleum is a non-replenishable resource, becoming an exploitable asset only after discovery. Since exploitation decreases both magnitude of reserves underground and the ratio of reserves to annual production, exploration has to continue, both to replace withdrawals due to exploitation and to increase inventories for increasing production in future. This implies that organised operations have to be undertaken to discover more and more hydrocarbon reserves and thereafter to exploit them. It must be remembered that the days of easy to detect prospects and exploration in accessible areas are over. Exploration now has to be oriented for difficult to detect subtle and stratigraphic traps. **Exploration is also to be conducted**

now onwards in deeper waters and in harsher and hostile environments.

24 New techniques and new concepts are being continuously developed and used in the search for non-conventional traps. For example, non-marine source of oil is one of the new thoughts in global oil venture. Therefore, our exploration strategy has also to be suitably reoriented and constantly updated. I am sure that geoscientists may take imaginative flights into fantasy, borrow from unproven theories and believe that more "Bombay Highs" must exist. This would necessitate, not only increasing the quantum of efforts but also improving the degree of sophistication, in order to obtain the greatest amount of oil at the least cost in the shortest period of time.

25 Sick Well & Enhanced Recovery Techniques : A more vigorous drive is required to exploit the oil reserves by resorting to enhanced recovery techniques. Restoring sick wells to health is another area which needs a big push to increase the oil recovery as well as to upgrade oil reserves.

26 **Development of Projects :** The development of projects and necessary corresponding **infrastructure** needs would merit serious thought to avoid sporadic and slip-shod growth resulting in lack of control, extra cost and built-in inherent inefficiency. Business decisions right at the start of the project will have to be taken to acquire land, etc. so that the development can take place in a coordinated manner. **Once we hit the oil in a non-producing area -** *a* **well coordinated drill will have to be laid down.**

27 **Participative Culture :** A participative culture has to be ensured so that the ideas of our young personnel are made use of in the best interests of the organisation. The younger colleagues will have to be given independent responsibility to enable them to develop faster. I have no doubt that our young engineers/executives/scientists and technicians will rise to the occasion if given due responsibility.

28 **Support from Headquarters :** The staff at various headquarters will have to make sure that the support is given to the people in the field so that the latter do not have to look over their shoulders. The attitudes of staff services will have **to be highly supportive.**

29 **Welfare Measures :** The welfare measures need equal attention to make sure that individual problems are tackled without any loss of time. It would be desirable that inventory of personal grievances is maintained at all Projects so that the same can be examined by the senior executives visiting from the Commission's Headquarters. I hardly need mention that it should receive the personal attention of all executives. All possible measures should be taken to improve the education and medical facilities further and accelerate the construction of houses, etc. I hardly need over-emphasise the importance of discipline which under no circumstances **will be allowed to be compromised.**

30 **Self Reliance :** Self reliance is our motto. At present many jobs are being carried out by foreign contractors particularly in the offshore operation. It would be necessary to develop in-house capabilities to carry out studies for various consultancy jobs presently being **undertaken by foreign consultancy firms.** Efforts have to be made so that we develop these in-house capabilities either independently or in collaboration with institutions and agencies within the country. It would be necessary to spot out such institutions/agencies within the country where basic facilities exist which can be developed to take up such work. A very long-term perspective has to be developed. Indigenous capabilities for technology and equipment have to be developed as a matter of great urgency.

31 **My expectations are very high and would demand highest standard of performance**.

32 We have to work with mutual confidence and respect as a good team. As a result of my interaction with a large cross section of our people, I find each one of you is looking with great expectation to a very bright future. **We have to convert this hope into a reality by creating culture of urgency to maintain targets and schedule**.

S.P. WAHI

Appendix – B

THE TIMES OF INDIA

...ember 1, 2000 • Capital   Bennett, Coleman & Co., Ltd.   40 pages with Ascent – 6 ps

GOVERNMENT DISREGARDS CAG AND ESTIMATES COMMITTEE

Oil field deals: Bad when in opposition, good when in power

By Rajesh Ramachandran
The Times of India News Service

NEW DELHI: When the Supreme Court last week dismissed a petition seeking a fresh investigation into the controversial Mukta-Panna oil deal, its decision ran counter to the findings of two other constitutional bodies; the Comptroller and Auditor General of India and the Estimates Committee of Parliament. Both bodies had pointed to severe irregularities in the case concerning the lease of lucrative oil fields to Reliance and Enron by the Narasimha Rao government in 1995.

In fact, the cross-party Estimates Committee, which had criticised the Mukta-Panna deal in its 1998-99 report, has dismissed the government's explanation of why oil fields discovered and developed by the Oil and Natural Gas Corporation (ONGC) were handed over to private players in the first place.

In its 1999-2000 report released earlier this year, the Committee has reiterated its criticisms and rejected the Centre's action taken report'.

Undeterred by the Committee report, however, the Vajpayee government is apparently finalising the production sharing contract for 12 more oil fields discovered and developed by ONGC. These include the medium-sized Ratna and R-series fields. Though Essar's contract was awarded a few years ago, the Dowe Gowda and Gujral governments went slow on privatisation because the CAG had strongly indicted the earlier contracts, including Mukta-Panna.

Ironically, finance minister Yashwant Sinha had, as an opposition leader, taken a stand sharply at variance with what his ministry and government are taking today. On June 27, 1997, he had said: "There should be a CBI inquiry to probe leasing of all oil fields to private parties, taking

- After Mukta-Panna, other oil fields will go to private hands, the Govt insists
- In this, the Govt has defied the estimates committee recommendations
- CAG too had pointed out irregularities in the Mukta-Panna award
- Surprisingly, when BJP was in the opposition, it wanted a CBI probe into all oil field contracts
- Now, the NDA Govt is going ahead with deals for other oil fields as well

all related aspects such as pricing of oil, cost sharing, actual current production, etc. There should be a re-assessment of oil reserves in all the fields leased to private companies, (and an) enquiry into the circumstances in which permission was granted to senior government and PSU officials to join private sector companies in fields directly related to their area of activity."

Three years later, when the Mukta-Panna case came up for hearing, the BJP-led government supported the deal. Interestingly, two Rajya Sabha members, one from the ruling party and the other from the Opposition, defended Reliance and Enron in the earlier stages of litigation: Kapil Sibal of the Congress, and the present law minister Arun Jaitley.

The Estimates Committee report of 1998-99 had stated: "The Committee (was) informed (by the government) that besides (the) rate of return not being attractive for development of above fields, the major foreign exchange crisis in 1990 was another factor for – awarding them to foreign private companies. The Committee, however, feels that (by) the time of awarding of Panna and Mukta fields in 1995 there was hardly any foreign exchange problem in the country..." In 1997, the CAG had also concluded that the privatisation of Panna and Mukta had resulted in huge losses to the government and ONGC.

The Committee opposed the handing over of the fields to private operators. The government's ATR, in defiance of the Committee, said it would continue with the privatisation of oil fields. And the Committee now refuses to accept the government's stand.

■ *Tomorrow: The Supreme Court vs the CAG on the Mukta-Panna deal*

Appendix – C

Relying only on oil could be slippery

By S. P. Wahi

(Financial Exprees: 16.09.1990)

IN the absence of an integrated national energy plan, the development of various energy sources in a balanced manner has suffered. Adequate attention has not been paid to work out strategies for conservation and efficient use of energy. Dependence on oil to meet the growing energy needs has further increased, making the economic growth very vulnerable to uncertain world oil supply and price situation.

The remarkable success of the ONGC in the eighties to increase the production of oil and gas, has encouraged the planners, to adopt the soft option of using oil as the swing fuel to meet the shortage and indifferent supply of other energy sources, i.e. electric power and coal. The use of standby generating sets by industry, agriculture and even in the domestic sector is a case in point. The availability of indigenous crude at $10 a barrel to the refineries has further made the planners complacent and no serious effort has been made to manage the oil demand effectively.

The recent Middle East crisis has again brought a sharp focus on the demand management, and efficient use of energy, in particular crude oil and its products. To produce a dollar of real GNP the industrial economies now use 40% less oil than they did in 1973. This has been possible by switching over to alternative fuels, effective conservation measures and use of energy-efficient equipment. Deliberate demand management strategies have to be formulated to discourage consumption of oil and not by hasty fire-fighting action.

Oil will continue to be a major and critical source of energy. The indigenous oil availability has increased from 0.25 million tonnes in 1947 to about 34.56 (taking oil equivalent of gas into account, the total production is 50) million tonnes in 1989-90, resulting in self-sufficiency of over 66 per cent. But the increase in demand is outstripping the increase in indigenous availability. What, then, should be the strategy to achieve self reliance in oil? It would be essential, first, to appreciate the oil game.

Oil exploration is a game of chance. It is a high-risk and possibly a high-reward business. There is no proven method of direct detection of hydrocarbons. The only definite means of establishing hydrocarbons is through drilling after detailed scientific surveys. A question which is generally asked is, as to why more Bombay Highs have not been discovered. It is well known that discovery of giant fields is a cyclic phenomenon closely related to technological breakthroughs and new concepts in geo-sciences. About 51 per cent of the recoverable hydrocarbon resources of the world other than in the centrally controlled economies are located in 49 giant fields while the balance 49 per cent are located in more than 30,000 medium and small fields. There are limited possibilities of finding large fields except in virgen areas still to be explored. As long as we are net importers of oil, any find big or small would be of interest if not today at least tomorrow. In the Sixth Plan the Share of oil supply of Bombay High was 75%, which reduced to about 60% in the Seventh Plan and would come down to 40% in the Eighth Plan. This indicates the rapid discovery and development of other fields.

Oil exploration is also a game of perservance. Oil was discovered in the Siberian basin of the USSR after 40 years of oil search and drilling of more than 90 days wells. Today this basin in producing about 250 million tonnes a year. Nearer home, experience in the Krishna-Godavari, Cauvery basins would illustrate the point further. Till date, approximately over Rs 1000 crores have been invested in the oil search lasting over 30 years. The recent discoveries of oil and gas in these basins should enable the investment to be recovered in the next few years and handsome profits thereafter. This business calls for "tough minded leadership" with an optimistic approach, a vision, creative thinking and ability to take calculated risks.

India has a reasonably strong hydrocarbon resource base. The prognosticated hydrocarbon resources are approximately 20 billion tonnes. This does not take into account the sedimentary basins offshore beyond 200 meters water depth. It is estimated that the hydrocarbon resource base in water depth beyond 200 metres, may be more than 5 billion tonnes. So far only about 5 Billion tonnes have been converted into geological in-place reserves through exploratory drilling. It is also fairly well known that the prognosticated resources, generally though not always, increase with time as more and more data is generated. This indicates the potential available for finding more oil and gas.

The exploratory drilling success ratio so far has been better than the world average. Exploration now, however, is moving into geologically complex and logistically difficult areas which would call for innovative technological inputs and cost effective measures. ONGC has a strong R&D base and most effective cost control systems to meet the future challenges.

Historically the planning in this country is based on five years. Taking into account the nature of the oil game it is imperative that oil exploration strategies are based on long term plans, so that the investment is incremental and not cyclic. The ONGC had prepared a long term conceptual plan up to the year 2004-2005 in 1981-82. This has resulted in a progressive increase in the plan outlays. In 1980-81 the plan outlay was Rs 468 crores, whereas in 1989-90 it was Rs 2803 crores. There has so far been no resource constraint as the plans have been financed through internal generation and the foreign exchange requirement gaps have been made up through commercial borrowings from the international market.

What, however, has not been possible sometime is to utilise the funds in tune with the plans, because of the multifarious controls

Is patrol rationing around the corner?

A solar heating device

exercised by the Government. In fact the performance of the oil sector would have been even better if the public sector was run on business lines by the Government.

This calls for a major restructuring of the public sector, and in particular the oil sector, to make sure that the companies are managed by professionals through the formation of holding companies.

To achieve fair amount of self-sufficiency in oil, a number of measures, some of which will be discussed in the succeeding paragraphs, need to be taken.

The ONGC's long-term plan up to 2004-2005 AD has projected a production potential of 120 million tonnes (55 MT OIL + 65 MT oil equivalent of gas). This plan is under updating to the year 2004/5

Taking the oil production possibilities upto the year 2014/5, a National Integrated Energy Plan should be developed, so that other sources (nuclear, hydro, coal and gas-based thermal power) of energy including unconvertial sources are developed to a time schedule to meet the growing energy needs. Oil should be used only for critical uses where oil replacement is impossible. Unfortunately today all demand projections are based on easy availability of oil and not taking into account the progressive increase in the availability of gas. The objective should be to move away from dependences on oil by the year 2014-5. This plan should also clearly lay down strategies to be followed for the conservation and efficient use of energy with due incentives. Lessons should be learnt from the experience of highly industrialised economies, the way some of them have managed their energy needs even without any hydrocarbon resources base.

The effective conservation measures, efficient use of energy, switching over to 'alternative fuels' after the 1973 and 79 oil shocks should be models for adaptation by us.

The exploration activity should be further intensified, even by opening up onland category III & IV basins to foreign oil companies including as joint ventures. Exploration activity in deeper waters beyond 200 metres water depth should be accelerated. Like other international oil companies, ONGC should accelerate exploration activities in foreign basins to have control over oil and generate foreign exchange. The recovery of oil from old fields should be increased through the use of secondary and tertiary methods of recovery. The upgradations of reserves of the old fields through the use of latest technologies and geological thinking should be followed up with great vigour.

The national gas grid proposal of the ONGC should be implemented. Linkages with the countries on the East and West and export/import of LNG through offshore terminals should be integrated with gas grid. This would enable, wide use of gas including in the domestic sector, to substitute liquid hydrocarbons.

The use of compressed natural gas in the transport sector should be ensured as a matter of great urgency to reduce the pressure on Liquid Hydrocarbons. We can draw on the experience of many countries, apart from the successful experimentation done by ONGC. 85% of the liquid hydrocarbons are being used in the transport and domestic sectors. The efficient use of energy and substitution by gas in these two sectors alone can reduce demand for oil substantially.

Approval of projects connected with zero gas flaring should be within the powers of oil companies. This would prevent national loss. (In Bombay Offshore alone almost Rs 600 crores are being lost every year). Revenue loss to the oil companies, and would ensure easy availability of gas.

To tap the large coal reserves (Almost equal to total prognosticated oil reserves) located at depths in oil fields, not minerable by conventional methods, the In-situe Coal Gasification Project of ONGC should be given more thrust for implementation on priority. This would make available in abundance another energy source for some applications.

Research and development work on non-conventional sources particularly solar being undertaken by many organisations/national bodies should be kept under focus for collaborative research where possible.

(Col. S. P. Wahi is a former Chairman of ONGC.)

Appendix – D

September 1987

REPORTER
THE CORPORATE JOURNAL OF THE OIL AND NATURAL GAS COMMISSION

CHAIRMAN'S CALL FOR INTENSIFYING EXPLORATORY EFFORTS

Since Independence, India has made spectacular progress in all areas of activities. The Oil & Natural Gas Commission since its inception has also made remarkable contributions to the country's economic development. Delivering the Independence Day address at the ONGC Stadium in Dehra Dun, Dr. S.P. Wahi said the country is passing through a critical phase of socio-economic transition-As proud citizens of this great country it should be our endeavour to shed all parochial and communal feelings and strive to achieve national integration to build a strong and economically self-reliant India.

Outlining ONGC's enviable record of growth and performance in the last few years, the Chairman said 1986-87 was the year of maximum performance of the organisation. It once again emerged as the number one profit making company with a record net profit of Rs. 1484 crores and a highest ever dividend declared at Rs. 35.99 crores with contribution to the exchequer of Rs. 3139 crores. The net worth has improved to Rs. 5913 crores in 1986-87. It has been declared as the eleventh largest profit making company in the world.

There have been 16 new discoveries since April 1986. Significant among these are GS-16-2 in Godavari Offshore, CA-CD Structure in Daman Offshore, Agartala Dome in Tripura, Namti in Assam etc. On August 10-11, two more discoveries were made in Bombay Offshore. One was gas and the other of oil, declared the Chairman amidst applause.

During the first two years of the Seventh Plan itself, 411.19 MMT of geological reserves have been added as compared to 894 MMT in the entire Sixth Plan. Krishna Godavari and Cauvery basin are emerging as promising new hydrocarbon bearing areas in the country.

A deliberate strategy to improve productivity in all areas of operations, cost control and greater emphasis on high technology areas and farming out low technology areas, has been adopted. Development of new oil and gas strikes have been given highest priority by mobilising resources in line with our strategy of reinforcing success.

A new dimension to the indigenisation effort by ONGC is the setting up of Joint Working Group with the Confederation of Engineering Industry. The foreign exchange outflow in relation to plan expenditure has been brought down from 69% in 1980-81 to 41% in 1986-87. In line with the open-door concept, Advisory Councils have already been set up in the areas of exploration strategy and human resource development. It is proposed to set up an Advisory Council on management and policy matters. It would include prominent labour leaders, educationists and experts drawn from the field of management and media. The objective is to strengthen the participative culture which is already prevailing in the organisation.

Dr S.P. Wahi taking the salute at the Independence Day celebrations.

With a view to establish recoverable hydrocarbon reserves at an accelerated pace, ONGC is launching a programme for exploration in deeper waters and also to explore in basins abroad for which negotiations have been initiated for operating in Basins of Vietnam and Tanzania.

Describing the significant role natural gas plays in the country's energy balance, the Chairman said we have added one billion cubic metres of natural gas reserves in the last 6 years and the utilisation has increased from 40% in 1980-81 to over 75% now. Large reserves of gas have been discovered in Tripura, Krishna-Godavari and Western Offshore and also to some extent in Rajasthan, Andamans and

Appendix – E

Hon'ble President visits ONGC R&D Institutes at Dehradun

The Hon'ble President, Mr. R. Venkataraman, while visiting ONGC Research & Development Institutes at Dehradun expressed his satisfaction at the excellent performance of ONGC in various fields of activities and also congratulated everyone in the organisation for achieving the desired results.

Col. S.P. Wahi, Chairman, presented an overview of the performance of ONGC in the context of its missions and objectives to the President. These achievements are all the more remarkable said the Chairman since they have come about in an environment packed with risk and danger. Col. Wahi pointed out the role of various innovative and planned strategies in this exemplary performance of ONGC. He also highlighted the challenges ahead for future growth and the steps required to be taken in this regard. This was followed by a presentation from Dr. S. Ramanathan, Member (P) on the comparative performance of the 6th and 7th plan and the projections for the 8th plan. The President wanted to know whether survey in the country has been completed. The Chairman and Vice Chairman appraised the President that the broad survey has been carried out in most of the sedimentary basins and that the exploration survey is an on going process. With the availablity of improved technology and more sophisticated survey techniques it would be possible to work in areas which earlier could not have been delineated. The President also evinced interest on the strategies of aeromagnetic survey. He was informed that the surveys have been completed in most parts of the country. Answering the President's query about the self-sufficiency in oil, the Minister of State for Petroleum and Natural Gas, Mr. Brahm Dutt, explained that self-sufficiency is a relative term and any action required in this area is deeply related to demand management. More and more substitutes of oil and gas, the usage of gas as an alternate fuel and other measures to increase fuel efficiency would contribute to reach this goal. Earlier on President visited the Institute of Drilling Technology, GEOPIC, KDM IPE and was taken around the museums and various labs of these R&D Institutes.

Some of the latest achievements that caught everyone's eyes here included the technique developed to interpret seismic

The President records his impressions in the visitors book at KDMIPE

data. Under this technique, date volume acquired from the sub-surface is taken as a data cube. This data cube is sliced along three orthogonal axis to verify data useful in hydrocarbon prospecting. This technique has been developed for the first time in India. Later briefing the President at the Conference hall, Col. Wahi said that efforts were on to develop a technology for recovering basement oil of the Bombay Offshore wells. The technique when developed is expected to double the life of these wells. Referring to the vast amount of timber being felled for fuel purposes, the President Mr. Venkataraman urged ONGC to expeditiously provide LPG as a cheap alternate fuel for the poor people of this country. Col. Wahi advocated for more autonomy for the Public Sector so that the plans could be executed at a faster pace and project cost escalation avoided due to the red tapism in the Government's machinery. Col. Wahi's views were shared by Mr. Brahm Dutt and Mr. Gulab Singh, UP Minister for Tourism who were also present.

Impressed by the achievements of ONGC, the President urged the scientists to disseminate works of social importance among the rural population so that the nation might benefit as a whole. The President also visited the Forest Research Institute (FRI) at Dehradun and met delegations from the public at the 'Ashiana' (President's estate) in Dehradun.

Reporter: January-March 1989

Appendix – F

President lauds ONGC's achievements

The President of India, Giani Zail Singh visited ONGC on June 11, 1985, where he was accorded a warm welcome. Col. S.P. Wahi formally welcomed the President at the conference room in KDMIPE. In a brief speech, he told the President about the tremendous growth in national oil production from a meagre 0.25 million tonnes at the time of Independence to over 29 million tonnes today.

ONGC registered a revenue income of Rs. 3,987 crores (1984-85) and a pre-tax profit of Rs. 1,768.07 crores. It contributed Rs. 2,092.22 crores to the national exchequer during 1984-85. Col. Wahi said that ONGC's performance during the Sixth Plan has been largely responsible in bringing India to a 73% level of self-sufficiency in oil.

Col. S.P. Wahi also apprised the President of the specific initiative taken by ONGC, especially in areas of long-term planning, restructuring of the organisation, intensification of training and development, R&D activities, and drive towards computerisation not only to cover technological needs, but for every business application, including information management. He informed the President that man management, environment management and safety are being given high priority in ONGC. Community service through adoption of villages has also been initiated by ONGC Ladies' Clubs.

President Zail Singh was conducted through the Institute of Drilling Technology, the Keshava Deva Malaviya Institute of Petroleum Exploration, the Computer Centre, the Training Centre and various laboratories at the ONGC. He was also apprised of the facilities for intensifying exploration and production efforts.

In an address at the KDMIPE conference room, the President complimented ONGC on its spectacular achievements. He said "The nation is grateful to ONGC for its contribution." The huge revenue contributed by it to the State exchequer and generation of internal resources are indicative of ONGC's role in the country. "We are proud of our scientists, engineers and other personnel for their dedication and commitment to national objectives." He said "Being a poor country, the people cannot be rewarded in the same measure, but the good work done by them would go a long way in building a strong India, a dream that had been cherished by Pandit Jawahar Lal Nehru, who visualised the need for a strong technological base in achieving self reliance."

The President expressed his pleasure at visiting the various ONGC Institutes and satisfaction with the task that the Commission is performing. He added that he was glad to see ONGC equipping itself with the best technology in the world. "It is a step in the right direction", he commented.

President Zail Singh felt that the objective of the Commission to use energy effectively and to promote development of alternate sources of energy, is indicative of ONGC's selfless commitment in the national interest.

The President concluded his address by reiterating his immense satisfaction with ONGC's performance. He said that in a vast country like India, ONGC through its wide areas of operation, contributed a great deal towards building the spirit of national integration. As also in bringing prosperity and upgrading the quality of life of the common man.

He said, "Greater challenges are ahead. But I am confident that the oil men of ONGC are geared to meet all of them."

Col. S.P. Wahi presents a memento to the President, on behalf of ONGC.

President Giani Zail Singh adds his comments to the Visitor's Book at the I.D.T.

Appendix – G

Performance Charts

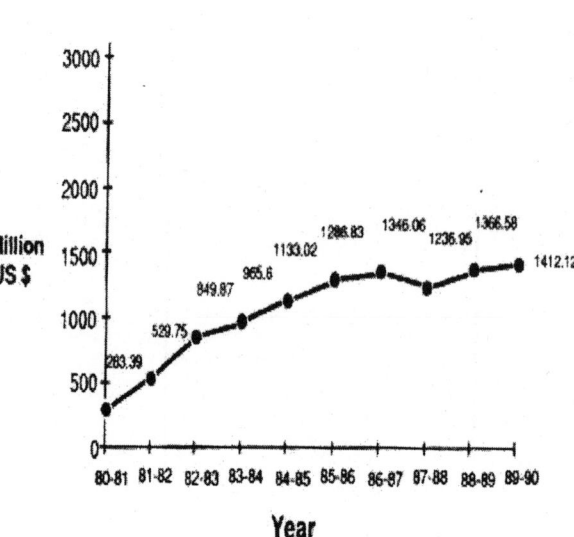

Source: Ministry of Planning & Programme Implementation, India

Annexure III

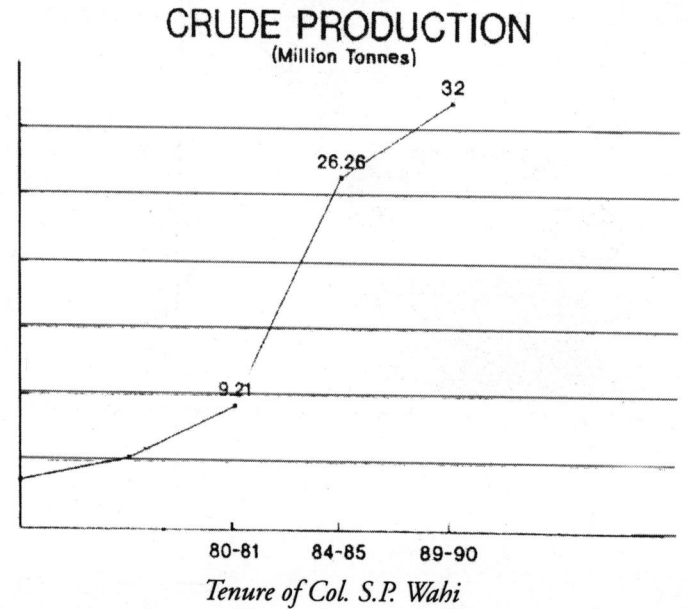

Tenure of Col. S.P. Wahi

Source: ONGC, India

Annexure IV

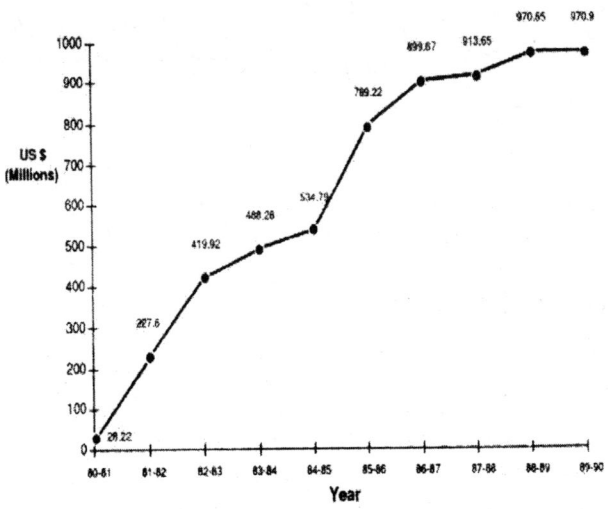

Performance During the Tenure of Col. S.P. Wahi

Appendix – H

Certain Statistical Comparative Figures to Mark the Growth of ONGC between 1980-81 and 1989-90.

(Tenure of Col. S.P. Wahi)

Parameter	1980-81	1989-90	Remarks
No. of rigs deployed:			A significant number of on-shore rigs were indigenously manufactured.
On-shore	34	116	
Off-shore	7	33	
Total	41	149	
No. of wells drilled:			A peak 146 wells were drilled in 1988-89
On-shore	72	491	
Off-shore	32	134	
Balance of recoverable reserves in MMT	366	726	Includes the figures of Oil India Ltd.
(Of which)			
Gujarat	52.73	155.12	Wholly ONGC
Off-shore	230.95	430.88	Wholly ONGC
Assam	82.65	140.24	Includes figures of OIL which however remained more or less static during this period.
Oil Production in MMT			In addition, production from Tamil Nadu and Andhra Pradesh commenced in 1986-87
On-shore	4.231	9.667	
Off-shore	4.985	21.716	
Of the above, production for Gujarat	3.808	6.313	
Gas-Reserves (in billion cubic metres)			Includes figures of OIL which however remained more or less static.
Gujarat	16.39	85.59	
Assam	63.53	135.47	
Off-shore	270.96	447.47	
Off-shore Supply Vessels indigenous	Practically none indigenous	67 of which 33 were owned by ONGC and the rest by Pvt. Companies, including 10 by Shipping Corporation of India.	Most of the supply vessels in 1981 were Foreign flag. In 1989, 99% were Indian flag.

Source: *"Indian Petroleum and Natural Gas Statistics"* issued by the Ministry of Petroleum and Natural Gas (1992-93)

Important Deductions

- Growth in exploration/Production is indicated by the deployment of rigs and wells drilled.
- Results-indicated by addition to the recoverable reserves and production per year.
- Major thrust in exploration resulted in Krishna Godavri, Cauvery Basins (Andhra Pradesh & Tamil Nadu), Rajasthan and Tripura becoming producing areas – a major contribution of the period.

- ONGC was getting only administered price which ranged around 8 dollars a barrel irrespective of International price during this period, which ranged upto 38 dollars a barrel.

It is only after the discovered fields of ONGC were gifted to the private sector, the international price was given to the indigenous producers. For sometime even then ONGC continued to get less than the price given to the private sector.

Appendix – I

Petroleum - Past, Present and Future

Dr. S. P. Wahi

One objective of energy planning is to ensure that the required energy is made available at reasonable cost to meet the growing demands of the economic development. In India, although the energy consumption has doubled in the last 10 years, but the per capita consumption is only one-eighth of the world's average. Even among developing countries, India stands pretty low when per capita consumption is compared. The demand for energy will increase with economic development and population growth. Meeting the energy demand is therefore going to be the greatest challenge in the coming years.

Energy Mix

India's energy balance constitutes 60% of commercial energy and balance 40% contributed by non-commercial energy sectors. The consumption of commercial energy over the past three decades has maintained an average annual growth rate of 6-7%, whereas the overall economy, an average annual growth of 3.5% per annum. The emphasis now will have to be achieving higher levels of economic growth with lesser consumption of energy. Improvement in energy efficiency is the key for sustaining desired levels of economic growth.

Petroleum has revolutionised our life. As a source of energy for domestic, industrial and transport sectors and as feedstock for agricultural chemical and other industries, it has become vital for economic development. It is also a necessity for individual households providing fuel for lighting, cooking and space heating. No wonder Pandit Nehru declared in Parliament on May 26, 1956. "Oil is of vast importance in the world today. A country that does not produce its own oil is in a weak position. From the point of view of defence, the absence of oil is a fatal weakness".

In spite of the International expert opinion that there was no more oil to be found in the country, it was the vision of Pandit Nehru and the perseverance of late K.D. Malaviya that necessary measures were taken for the exploration of oil.

Challenges

Oil will continue to be a major and critical source of energy in the world. In India, it is anticipated to meet about two-fifths of the commercial energy consumption by the end of the century. The indigenous oil availability has increased from 0.25 million tonnes in 1947 to about 30.36 million tonnes in 1987-88, resulting in self-sufficiency of over 62 percent, but the increase in demand is outstripping the increase in indigenous availability. What then are the challenges and the strategies to be followed to achieve self-reliance in oil ? It would be essential first to appreciate the oil game.

Oil exploration is a game of chance and is considered to be the greatest scientific gamble. It is a high-risk and possibly a high-reward business. There is no proven method of direct detection of hydrocarbons. The only definite means of establishing hydrocarbons is through drilling after detailed scientific surveys. A question which is generally asked is – why more Bombay Highs have not been found. Oil is not manufactured but is located. It is also well known that discovery of giant field is a cyclic phenomenon, closely related to technological breakthroughs and new concepts of geosciences.

About 51 per cent of the recoverable hydrocarbon resources of the world other than in the centrally controlled economies are located in 49 giant fields while the balance 49 percent are located in more than 30,000 medium and small fields. There are limited possibilities of finding large fields except in virgin areas still to be explored. As long as we are net importers of oil, any find big or small would be of interest if not today at least tomorrow.

Oil exploration is also a game of perseverance. Oil was discovered in the Siberian basin of the USSR after 40 years of oil search and drilling of more than 90 dry wells. Today, this basin is producing almost 250 million tonnes a year. Nearer home, experience in the Krishna-Godavari basin would illustrate the point further. Till date, approximately Rs. 6200 million have been invested in the oil search lasting over ten years and without any revenue.

The recent discoveries of oil and gas in this basin should enable the investment to be recovered in the next seven years and handsome profits thereafter. This business calls for 'tough minded leadership' with an optimistic approach, a vision, creative thinking and ability to take calculated risks.

Hydrocarbon Resource Base

India has a reasonably strong hydrocarbon resource base. The prognosticated hydrocarbon resources are approximately 18 billion tonnes. This does not take into account the sedimentary basins offshore beyond 200 metres water depth. It is estimated that the hydrocarbons resource base in water depths beyond 200 metres may be more than 5 billion tonnes. So far only 5 billion tonnes of prognosticated resources have been converted into geological in-place reserves through exploratory drilling. It is also fairly well known that the prognosticated resources, generally though not always increase with time, as more and more data is made available. This indicates the magnitude of effort that is still required to establish the balance reserves.

Production Potential 2004/5

The drilling density in India is still very low compared with the world average. As against the world average of 95 wells in 10,000 sq.km, in India it is only 12 wells. The exploratory drilling success ratio so far has been better than the world average. We are now, however, moving into geologically complex and logistically difficult areas which would call for innovative technological inputs and cost effective measures. A long-term assessment of the likely availability of oil/gas from Indian sedimentary basins was done through modeling exercises based on the hydrocarbon resource base in 1981-82 and updated in 1986. The most optimistic scenario is around 120 million tonnes (65 oil and 55 gas) of oil and oil equivalent of gas by the year 2004-05.

Growth Production

There has been continuous increase in the production of oil and gas is the country. The growth of reserves has been encouraging to sustain higher levels of production. The geological in-place reserves of oil and oil equivalent of gas as on 1st January, 1988 stand at 5 billion tonnes (3.726 oil and 1.274 gas). The production of oil and oil equivalent of gas in the Seventh Plan was an increase of 67% over the Sixth Plan. The likely target of the Eighth Plan is 338 million tonnes which will be 189% more than the Sixth Plan. The growth of gas production in the Eighth Plan is going to be phenomenal – 808% over the Sixth Plan. The oil production in the terminal year of the Eighth Plan is projected to be over 48 million tonnes as compared to 34.31 million tonnes to the end of the Seventh Plan – an increase of 42%.

Source: Urja, January 1989

Appendix – J

OPEC (Organisation of Oil Producing and Exporting Countries)

It is important to understand the critical role played by OPEC since its formation, to influence the balance between supply and demand of OIL and its price. The volatility in oil prices has been a source of serious economic concern for the oil importing countries such as USA, Japan, China and India. It is essential to understand the historical perspective, leading to the formation of OPEC.

OPEC is an organisation of twelve oil producing and exporting countries from Africa (Gabon, the Socialist People's Libyan Arab Jamahiriya and Nigeria); Asia (Indonesia); the Middle East (the Islamic Republic of Iran, Iraq, Kuwait, Qatar, Saudi Arabia and the United Arab Emirates); and Latin America (Venezuela). Gabon has since withdrawn from OPEC (BP Statistical Review 2003). OPEC was born of protest against the high-handedness of the major international oil companies, the so-called "Seven Sisters". Before the advent of OPEC, these seven multinational oil companies- EXXON, Texaco, Royal Dutch/Shell, Mobil, Gulf British Petroleum, Standard Oil of California and Chevron had created "states-within-states" in the oil producing countries, controlling the amount of oil extracted, sold, to whom and at what price. On these matters of vital importance to the livelihood of the oil producing countries, the host governments were never consulted. They were paid a meager royalty, while the oil companies made vast profits and built huge financial empires for themselves at the expense of the producing countries.

This unsatisfactory state of affairs came to a head in 1959, when the oil companies unilaterally cut the posted price for Venezuelan crude by $0.50/b, and that for Middle Eastern by $ 0.18/b, followed, in August 1960 by further reductions in posted prices for Middle Eastern crude of between $ 0.1 and $0.14 per barrel. One month later, in September 1960, a meeting was held in Baghdad, which was attended by representatives of the Governments of Iran, Iraq, Kuwait, Saudi Arabia and Venezuela. That historic meeting ended on September 14 with the birth of OPEC.

OPEC Will Continue to Play a Major Role in the Oil Market

The role of OPEC in the world oil industry has been a significant one since the Organisation was formed in 1960, and that role is expected to expand in the decades ahead. OPEC produces about 40 per cent of the world's marketed crude oil. OPEC's marketed share should also rise over time. OPEC's forecast, which is in line with the forecasts of other institutions, shows that total world oil demand should reach about 100 million barrels per day (m b/d) by the year 2020, up 44 per cent from 1995. Total world gas demand should grow even more strongly and by the year 2020 should reach around 3,120m tonnes of oil equivalent, up 84 per cent from 1995.

OPEC's share of the oil market should reach roughly 50 per cent by the year 2020. Importantly, this would require OPEC to produce around 50m b/d of crude oil, almost double our current output. Although OPEC holds roughly 77 per cent of world proven crude oil reserves, this forecast growth in output represents a considerable challenge for a group of developing countries, as it will require billions of dollars for exploration and development.

OPEC is also an important and growing force in the world gas market. OPEC holds some 42 per cent of world proven gas reserves, yet it currently produces only 14 per cent or so of the world's requirements. OPEC Member Countries, particularly those in the Middle East, are among the world's largest repositories of natural gas. In contrast, the largest non OPEC gas reserves are to be found in the former Soviet Union, which does not have a geographical advantage in exporting its gas. Logic therefore dictates that OPEC has significant scope to substantially increase its production and share of the gas market over the medium to long term..

Source: An extract from Dr Shokri Ghanem's paper published in *OPEC Bulletin*, July 1998

An extract from *"OPEC AT A GLANCE"*

OPEC means different things to different people. It is to many, overwhelmingly in the Third World, a shining example of the economic power developing nations can wield to improve their conditions of life if they act in concert to exploit the natural resources with which they have been endowed. To others, almost without exception people from the industrialized countries the Organization is still, to a limited extent, an 'adversary' which held affluent societies to ransom by raising oil prices and whose continued existence constitutes a threat, either to 'national security' or to the so-called 'free markets'. What then is OPEC?

Appendix – K

Wedding bells in the distance: Tiny tots in fancy dress gear as brides & bridegrooms.

The inspiring force behind 'Sanghe Shakti'

In ONGC, the name Shobhana Wahi is synonymous with dynamism, for in her association with the Commission she has proved that one woman with vision and compassion can change an entire environment, not only within the organisational confines but for all women of India.

It is difficult to get Mrs. Wahi to speak of her achievements for her reply is always a modest rejoinder of how little she feels she has achieved in comparison with her hopes and aspirations for Indian womanhood.

Shobhana Wahi has left her unique and indelible imprint on all organisations with which she has been associated from the early days until now. In the Army wives welfare Association Mrs. Wahi played an active part in adult education programmes and rehabilitation of war widows. Even in this initial period of her awakening to succour the needy, Shobhana Wahi showed marked signs of the visionary zeal that has developed to be her hallmark. Later, at BHEL she introduced the idea of vocational training for the poor and destitute. Under her

Mrs Shobhana Wahi

dynamic tutelage the ladies of BHEL geared upto meet all challenges of changing the environment from one of despair for the poor to an era of hope and self fulfilment.

In ONGC, Mrs. Wahi felt the pulse of the Commission's greater family and decided to tap the vast unutilised woman power lying dormant in wives and dependents of ONGC employees. This was done by gradually enthusing and encouraging the ladies to come forward and unite to plumb not only their own potential but also foster this spirit in others. This was the beginning of a vast new era in the Commission through the activities of the Mahila Samiti. Firstly, the focus was on forwarding the 20 point programme initiated by our Late Prime Minister Mrs. Indira Gandhi and later carried on by our Prime Minister. Under this, Mrs. Wahi and her team organised rural upliftment programmes through adult education, medical facilities, eye camps, child care centres and installation of biogas plants. The intrinsic involvement of Shobhana Wahi was the key to success— "No effort was spared to ensure that efficiency coupled with compassion would mitigate the hardships faced by villagers."

Mrs. Wahi's tender love for children caused her to open a nursery school, the Shishu Vihar. She then envisaged a much more ambitious and far reaching plan for a Welfare Centre to train the poor and needy to help them take their right place in the economic development of our nation.

Mrs. Wahi personally supervised all the branches of ONGC Mahila Samiti throughout the country and her message of 'Sanghe Shakti' is a byword for all its members.

Mrs. Wahi's dream has come true with the opening of the ONGC Mahila Samiti Polytechnic for Young Women in 1987. The response was so overwhelming that she has now started, Hostel facilities thus opening the benefits of the Polytechnic to all women of India. Mrs. Wahi personally selected the courses and formulated syllabbi. Inspite of her busy schedule she rigorously monitors the performance of the students.

Mrs. Wahi's vision for the future envisages a multi-pronged strategy to further women's upliftment. In the rural programmes the focus will now be on formal primary and secondary education facilities, vocational guidance and village in-situ training programmes, promotion of village handicrafts, etc. The welfare centre activities are being rapidly expanded giving thrust to imparting economically profitable skills to uneducated women, simultaneously educating them. For the Polytechnic, Mrs. Wahi envisages the addition of courses in the latest fields like Mass communication, Management and Master's Degree in Social Welfare (MSW).

Mrs. Wahi's generosity of spirit always causes her to play down her role in this far-reaching social revolution. Her reply to all praise is a humble "Sanghe Shakti".

COUNTRY PROFILE
INDIA

ONGC makes its presence felt internationally

ONGC is making its presence felt in other parts of the world, as a part of its deliberate strategy towards internationalisation of its operations. It has signed a Memorandum of Understanding (MOU) with PTT Exploration and Production Company Ltd (PTTEP) — a subsidiary of the Petroleum Authority of Thailand. The MOU was signed by Col S P Wahi, Chairman, ONGC and Dr Tongchat Hongladaromp, President, PTTEP in the presence of the Prime Ministers of the two countries.

The Thai delegation visited a number of ONGC installations offshore and onshore, and the R&D and training institutes at Dehradun and Ahmedabad. The decision to seek ONGC's expertise had been taken only after it had ascertained ONGC's capabilities and expertise. The Thai petroleum authority in its earlier assessment visits to the country had expressed their deep satisfaction with the infrastructural facilities and capabilities of ONGC. They had been impressed by the comprehensiveness of the range of services that could be offered by a single company.

Starting from exploratory surveys and basin evaluation to Laboratory Studies and ultimate field development, ONGC was able to provide its expertise in all relevant areas of oil exploration to production.

Areas for possible cooperation with PTTEP:
• Joint exploration ventures in Thailand and Monitoring of Exploration Contracts.
• Oil/gas field development plans.
• Reservoir Modelling, Reservoir performance analysis, development of EOR Schemes for depleted fields.
• Basin Evaluation and analysis of the Thailand basins both onshore/offshore as a second opinion.

• Laboratory studies of all kinds viz: Palaeology, Palynology, Sedimentology, geochemistry, SEM, etc.
• Training of Thai engineers/specialists in R&D institutes of ONGC.
• Deputation of ONGC specialists on a short term basis based on requirements of PTT.

ONGC is presently executing three other overseas contracts worth US$2.06 million. While the contract presently under implementation in Abu Dhabi is at an advanced stage of execution and is likely to be completed this year, the contract with Petronas — the National oil company of Malaysia has been signed only recently.

In 1987, ONGC had entered into a consultancy contract for exploration with the Abu Dhabi National Oil Company (ADNOC). The entire work of this project is being done at the ONGC's premier R&D institute — Keshav Deva Malaviya Institute of Petroleum Exploration, Dehradun. ONGC has already received a payment of US$987,356 out of the total amount of US$1.43 million.

During January 1989, ONGC had also concluded an agreement worth US$627,724 with the National oil company of Malaysia — Petronas — for undertaking two consultancy projects. The two projects, namely: (1) Geohistory Analysis of Tembungo Area, and (2) Exploration of Waxy Dulang Crude, are to be carried out at the ONGC's Institute of Petroleum Exploration at Dehradun and the Institute of Reservoir Studies, Ahmedabad. Petronas has awarded these projects to ONGC after a Petronas task-force had visited several institutes of various oil companies in several countries, including ONGC. After being thoroughly impressed with the expertise of ONGC and its pertinent experience in basin evaluation and reservoir and production engineering, Petronas decided to award these projects to ONGC against very stiff competition. For technology transfer, Petronas people would be associated with ONGC personnel for short periods during key phases of both the projects. The projects would be completed in about two years. Petronas is also considering mutually beneficial cooperation with ONGC in areas such as Enhanced Oil Recovery, Drilling and Training of Petronas personnel in various areas of oil exploration and exploitation.

Meanwhile, ONGC has already commenced seismic survey work in Vietnam. Hydrocarbons India Limited (HIL), a subsidiary of ONGC, had entered into a petroleum production sharing contract offshore Vietnam in May 1988, for which ONGC competing independently amongst international bidders was awarded the initial seismic survey contract. The complete production sharing contract is for a period of 25 years, with an option with HIL for withdrawal after four years.

Schlumberger sets up fifth centre in India

THE ONGC-Schlumberger Wireline Research Centre was formally inaugurated by Shri Brahm Dutt, the Minister of State for Petroleum & Natural Gas, India, recently. The Research Centre is a joint venture between ONGC & Schlumberger — a leading international oil service company. This is the 5th R&D centre with which the Schlumberger is associated. The other four centres are in the USA (two), France and Japan.

The purpose of the centre is to conduct research into wireline well logging. Wireline well logging is the process by which geophysical instruments are lowered on electrical conductor cables ("wireline") down prospective oil wells, to survey the subsurface. Measurements are made with respect to depth or time ("logs") and are transmitted up the wireline to surface recording equipment.

The overall objective of the research centre is to apply the latest techniques in subsurface geophysical measurement, and related technologies, in the oilfields of India with the aim of improving efficiency in the exploration and exploitation of hydrocarbon reserves.

34 PETROMIN, JULY 1989

Appendix – M

DO No. 6/33/CH/86. Oil & Natural Gas Commission, Tel Bhavan, Dehradun-248 003, August 1, 1986

S.P. WAHI
CHAIRMAN

Dear

ONGC in the last five years has been able to achieve phenomenal growth with stability and continuous improvement in productivity. The statistics have been brought out in a booklet titled "ONGC's growth/Performance Indicators/Productivity Improvement" (This booklet is available in all the Regions and every executive must study the same). A large number of new initiatives have been taken, some of which are also recorded in this booklet. We can be proud of this performance. Our performance in the first year of the 7th Plan period has also been commendable.

2. We have still a long way to go to achieve perfection. I, therefore, wish to share some of my thoughts with you, to achieve the above objective.

3. The performance of ONGC is generally viewed with reference to following by the outside agencies:
   a) Turnover and profits
   b) Production of oil and gas and percentage of self-sufficiency achieved in the country
   c) New finds and addition to the geological in-place reserves and specifically increase in recoverable reserves
   d) Cost of production of oil/gas, cost of drilling, cycle speeds and cost of exploration
   e) Inventory holdings
   f) Technological self-reliance

4. For achievement of the above performance indicators, following factors are important:
   a) Long-term Planning
   b) Detailed Planning and Budgeting
   c) Human Resource Development
   d) Human Relations — man management leadership
   e) Innovative and creative activities
   f) Culture of excellence
   g) Reduction of manpower in Regional and Commission Headquarters
   h) Optimisation of Human and Material Resources
   i) Research and Development
   j) Information Technology
   k) Cost control and reduction
   l) Conservation and efficient use of energy
   m) Environment and safety
   n) Reduction in inventories
   o) Communication
   p) Entrepreneurial spirit and sense of belonging
   q) Modernisation of equipment, technology and knowledge
   r) Management through the government
5. I shall touch on some of the above, in the succeeding paragraphs. Some of the new policy decisions which have been taken recently to optimise our operations are as follows:
   a) Over the years, ONGC, like many other Public Sector companies, has brought under its control, **lowest to the highest technologies** resulting in inadequate management time to the high technology areas. It has, therefore, been decided to farm out low/medium technology areas by motivating our employees to retire and form cooperatives to render services to ONGC in such areas. The areas which have been presently identified are shot hole drilling, transport and maintenance. A few Shot Hole Drilling cooperatives have already been formed in the Western Region. This has improved the productivity manifold.

b) **A cooperative movement has been launched so that the wards of the deceased employees, retired employees and their wards could form cooperatives to supply services and materials to the ONGC**. A number of vocational centres are also being set up through Ladies Clubs.

c) Assistance is being given towards the formation of private sector companies to provide services to the ONGC so that the future purchase of capital equipment can be reduced. Some of the areas for which such actions are being taken are contract drilling on land and offshore, seismic data acquisition and processing, logging services, mud engineering and cementation services.

d) **Efforts are also being made to farm out Logistic support services for offshore operations like OSVs.**

e) **No additional maintenance and workshop facilities to be set up within ONGC.**

f) **No more transport equipment to be inducted.**

Manpower Optimisation

6. The above actions would ensure balanced growth of manpower. This, however, would not affect total job potential in the country as these services would be rendered by other agencies within the country.

7. One common factor contributing to the failure of enterprises, as noted from the study of such enterprises the world over is uncontrolled growth of manpower. The optimisation of manpower resources therefore merits highest priority. The manning norms must change with the introduction of new technologies, computerisation, better facilities at sites, improvement in logistics and communication. The 'tail to teeth' ratio must be kept very low.

8. There is surplus manpower in a number of areas of our activities. The Commission and Regional Headquarters are overstaffed. Drastic reduction measures have to be initiated. It is the duty of every executive to analyse this systematically and make sure that every individual is contributing to the organisation, commensurate with his position and compensation received by him from the organisation. This can only be achieved if the

tasks/results expected from each individual are quantified and high standards of performance are demanded. As leaders, we have the responsibility to ensure this.

9. To ensure growth, a large number of people have been upgraded in the last five years. **But we have to ensure that expertise on the 'jobs' is not diluted particularly in the areas of drilling and production installations**. This has to be ensured by upgrading the level of supervision on Rigs and on Production installations.

10. The Senior Executives must move forward for giving advice and assistance at the workplace rather than holding meetings in the offices and making productive time of those responsible for quantifiable results, getting lost by moving back to the Headquarters. Some of our Senior Executives who have followed this strategy have shown excellent results. Focus must remain on the man responsible for quantifiable results. He should not have "to look over his shoulders" for help.

Cost Control

11. Each one of us has to resolve to achieve perfection in our areas of operation both qualitatively and quantitatively. The cost control and cost reduction measures have to receive the highest priority. This can only be achieved if deliberate effort is made to economise and reduce expenditure in every possible manner. We must have **'Millionaires mentality'**. The work has to be done at a given cost and not at any cost.

12. The first task is to optimise equipment and human resources and improve productivity. The cost reduction will automatically be effected in case the quantifiable results improve. Therefore, operational efficiency in every discipline has to be of the highest order. The best **International Forms** should be **the target** for achievement. For this, data must be collected through study of literature, visits abroad and discussions with foreign contractors operating in India. Deliberate efforts should be made to collect data from other companies by personal liaison and also through Indians settled abroad. We should establish contacts with Indians settled abroad, operating in areas of our interest. They could be "offered company's hospitality on their private visits to India for interacting with our Scientists/

Engineers/Executives. This however will have to be done in line with the delegation of authority.

Use and Maintenance of Facilities

13. The care, maintenance and use of equipment and facilities, in particular 'article in use' merit a lot more attention than in the past. I find a large number of photocopying machines have been bought but most of them are out of operation and work continues to get done from outside. This I am only quoting as an illustration as a large number of other 'article in use' receiving similar treatment, which to say the least, is most unacceptable. The areas which call for strict cost reduction are as follows:
    a) Entertainment
    b) Taxis
    c) Travelling
    d) Telex/telegram/telephone
    e) Stationery, etc.
    f) Optimise use of 'article in use' such as typewriters, photo copying machines, computers etc.

14. The Entrepreneurial spirit is seen lacking at times, when I find that facilities hired are not being optimally utilised or similar owned facilities are either unutilised or partially utilised. This is apparent in the case of houses, warehouses, transport and expensive NDT, MSV and Offshore Supply Vessels. There have been cases when it has taken inordinately long time to commission the new equipment. This is a matter of serious concern.

15. The housekeeping, particularly at the drill sites, continues to be poor though in certain areas very significant improvements have taken place. The sanitation at the camp sites/drill sites is of paramount importance. The standard of cleanliness of the toilets and other common facilities merit positive improvement. Needless for me to mention, the efficiency of an organisation can be immediately assessed by looking at such facilities and housekeeping.

Administration

16. The despatch of papers/files from one office to the other and one station to the other, show lack of urgency, resulting in tremendous loss of time and creating embarrassing situation when cases are investigated either as a result of top management or the Government intervention. **A culture of urgency must prevail down the line.** This can only be created by setting a personal example by the executives.

17. The proper maintenance of the assets created is our duty. In particular, maintenance of housing colonies, dispensaries/ hospitals, parks, sports grounds, school buildings and other buildings and offices, merits very high priority attention. Regular inspections are a must to maintain the desired standards. We should not accept 'slipshod' work. The Zero Defect culture must prevail.

Civil Instruction

18. I regret that quality of civil construction leaves much room for improvement. The designs are not properly scrutinised and quality during construction is not controlled resulting in creation of buildings/ assets which would stand as 'monuments of inefficiency'. I would urge all the Civil Engineers to take necessary measures to improve the situation. The value engineering concept would bring about the desired cost reduction.

Decision-making

19. The decision-making at various levels continues to be slow. We continue to follow bureaucratic culture of 'speaking' on the files. A participative culture must prevail. Those involved in the decision-making should get together, discuss, finalise and record a common note. We should not stand on formalities or have false vanity and avoid going to each other's offices for this purpose.

Material Management

20. I have had to face very embarrassing questions from the Government with reference to the delays in handling of purchase/procurement cases. Needless for me to mention that

operational targets cannot be achieved unless materials/equipment are procured as a matter of great urgency, in time. The specifications have to be drafted with care to avoid innumerable problems during evaluation of tenders. The bid evaluation criteria also should be drawn with care, so that all prospective bidders get a fair deal. We have not only to be honest but look to be honest.

21. The equipment and stores on receipt are not taken on charge immediately, resulting in delay in finalisation of papers and payment to the contractors. A large number of stores continue to be in transit due to lack of seriousness in completing the paper formalities and delay in movement of stores and materials. A lot of money is being lost in warehousing at ports and demurrage.
22. It is a matter of concern that a large number of packages received from abroad years ago have not even been opened.

Condemned Materials

23. There is a lack of concern in disposing of condemned materials, stores and equipment resulting in blockage of useful storage space, delay in realising money through sale, and infructuous expenditure in maintenance, storage and accounting.

Inventory

24. The inventory is showing an alarmingly increasing trend in absolute terms which is most unfortunate and merits utmost care in indenting and procurement of spares, etc. The executives must visit the stores to get an impact of the high inventory holdings. The indents are being floated without much care and concern.
25. It is for consideration whether for spares a separate subsidiary company should be formed or I should be farmed out to warehouses to be established in the private sector.

Management of Contracts

26. The Management of contracts is another weak area. The contractors try to build up cases for extra time and money through letters which are either not replied or replied late making our position weak when the cases go for arbitration.

We must settle the disputes with the contractors there and then, rather than 'wishing them away' with the hope that they will automatically disappear. A lot of sense of urgency is required.

Defects —New Equipment

27. Full advantage must be taken of the warranty clause. The defects must be reported expeditiously. We must show the same concern as for our private possessions.

Creativity

28. The dynamism of an organisation is seen from the number of ideas and suggestions being put up through the innovative and creative minds of the people. I am afraid, we have to go a long way. The executives at various levels have been advised to create **'thinking groups'** by involving a cross section of the people to discuss problems and come out with new ideas. Such **'creative thinking groups'** enable an organisation to think of alternatives other than the methods being followed for various activities, for better results. It is quite likely that a number of initiatives/ideas thrown up and implemented are not being projected with the result that those responsible for the good work are not being suitably rewarded.

29. We need to experiment with new ideas and geological concepts to find more oil/gas at the earliest at the least cost. This is the call of the hour.

30. The organisation has adequate resources to reward those with new ideas, concepts, inventions and techniques to improve the efficiency.

Long-term Planning

31. The Impact of ONGC's long-term conceptual plan upto the year 2004-05, which was formulated in late 1981 has already been felt within the Commission and in the industry. There has been growth of the organisation and of the personnel. The industry has been able to take investment decisions keeping in view the long-term needs of the ONGC. A large number of companies have come up to supply equipment, materials and services. There has literally been an **Industrial Revolution.**

32. The long-term planning has to be a culture with us for every activity so that ad hoc decisions do not compromise the future growth. All short-term activities/plans must dovetail into long-term plans.
33. As advised in the past, each region must formulate its own long-term plan.

Exploration, Exploitation and Eor

34. We have had more than forty new strikes/finds in the last five years. The upgradation of category C_1 and C_2 reserves merits a real thrust. The new discoveries have to be put on production with least delay. The utilisation of gas deserves constant attention to reduce flaring. This can be brought about through new and unconventional ideas and aggressive marketing. **The optimum utilisation of gas can generate a lot of additional revenue and profits.**
35. EOR techniques have to be developed at an accelerated pace as we cannot afford to leave oil in the reservoirs in view of continuing growth in demand of oil.

Drilling

36. The drilling activities on which maximum money is being spent, have shown considerable improvement in the recent past. The performance of the Eastern Region has been extremely good last year as compared to the past performance. In this area we have to have a relook on the land drilling rigs to have a certain amount of rationalisation so that the rigs selected/deployed are in line with the tasks in the Regions/basins. The similarity of the rigs in each Region would ensure better maintenance and spare parts management. The introduction of more mobile rigs and hell-rigs has to be ensured as a matter of great urgency to achieve better cycle speeds and operational efficiency.
37. The rigs and associated equipment particularly the BOPs call for more rigorous maintenance and periodic inspection to have **high reliability and safe operations.**

Technical Services

38. There has been considerable improvement in the technical services, particularly in the area of logistics. The availability of material handling equipment has to improve further. The personnel in the Technical Services Group have to meet the requirements of the operational group **'at the place, at the time and at the cost required by them'.** In this lies the real pride of the Technical Services Department.

Communication

39. **Communication with the remotest sites with the Regional Headquarters and with the International Oil Companies has to be the main objective to be achieved before October, 1987. This will enable quick and positive support to the operations based on live data instead of carrying out post mortems on historical data.**

Computerisation

40. A massive campaign has been initiated to augment computerisation capacity for seismic data processing at Dehradun and in the regions. Computers are being introduced for all business applications including career planning and medical services. **Each one of us must train ourselves for the use of computers. The training institutes have been directed to spend some time on each course on computer orientation.**

Finance

41. The management of our financial resources deserves a lot more attention to get the best returns. The payments of the contractors/suppliers must be made in time as a matter of routine and our obligation. Similarly, our dues must be collected from customers with real concern and aggressive approach.

Management Studies/Audits

42. A systems approach is required to optimise every activity. A number of Management Studies have been undertaken by corporate MSG, resulting in substantial savings in OSV, Helicopter operations, warehouses, etc. This approach should

be followed at all levels. The executives must read the reports so far prepared to get an impact of the possibilities. Instructions are being issued, that copies of such reports should be available at least in the regional libraries.

Research and Development

43. R&D activities in the institutes should be relevant to the operations in the fields and to meet the immediate future needs. Most of the projects should be sponsored by the regions. The basic research should be farmed out to the educational institutions and other scientific organisations set up in the country. ONGC had organised sometime back a brainstorming session with the universities, IITs, etc. with very fruitful results. This is being followed up. Our R&D Institutes will work as profit centres. This has already brought about the desired cost, quality and time conscientiousness. We must have a relook at the R&D projects under progress.

Training and Development

44. I am afraid we have to go a long way in improving our services both as professionals and as administrators. In spite of many measures which have been taken in the recent past on training and development, establishment of libraries, the general awareness of the executives in areas other than their own continues to be low. We seem to follow the 'frog in the well policy'. Any one 'who ceases to improve ceases to be good'. Each one of us must spend some time for personal development. The desired facilities have been provided and can further be augmented on request. Reading habits have to be cultivated. **We should not be carried away by what we had learnt in the colleges and through experience. There is a knowledge and technological explosion around us and unless we keep abreast with the latest we would become obsolete — 'knowledge obsolescence is worse than equipment obsolescence'.**

45. I find during interaction in interviews that executives have very little awareness of the Material Management, Administration and Finance disciplines. As executives we have to manage resources — both human and materials, and therefore our

knowledge of these functions must improve. The officers of Finance and Personnel disciplines can only assist you but the real responsibility rests with the operational managers.

46. The annual reports of the Commission should be studied very carefully. As executives, you have to play the role of PR to the people outside the organisation and therefore important facts and figures about the Commission should be on your finger tips.

47. I have asked our Training Division to organise courses both on material management and finance for non-material and non-financial managers. The regions should also organise courses on the subjects.

48. As executives, it is our duty to take care of the training needs of the people who report to us. The training of every individual is important to enable him to perform his duties more efficiently. This is applicable to class III and IV staff as well. Such training programmes can be organised locally.

Career Planning

49. We have established a reasonable data-base which needs to be further perfected for career planning of individuals. The job rotation and training programmes should be organised for each individual for his movement up the 'ladder'. It is absolutely essential for each executive to have shouldered responsibility of operations in various disciplines before he can move upto the highest position in the organisation.

50. We have inducted a large number of young executives in the last four years and may be their utilisation has not been optimum, resulting in frustration. All concerned have to pay particular attention to the development of the young executives by keeping them overloaded with work. They are the future senior executives of the company and have to be groomed with care. No young executive should remain in the R&D institute, Commission and Regional Headquarters for more than three years.

51. As executives, our conduct, behaviour and discipline have to be of the highest order. We have to set a personal example. ONGC executives should 'stand out' by their looks and bearing

and for that, needless to mention, we must dress appropriate to the occasion. The culture of excellence in this area is extremely vital for the good name of the organisation.

Man Management

52. We have to show a lot more concern in dealing with our people. We have to be **fair, friendly and firm.** We must take care of personal problems of our people personally and not impersonally.

53. There have been a number of cases where individuals have remained under suspension for long time. Such cases have to be examined as a matter of great urgency. The settlement of dues of the families of the deceased employees should be a matter of great concern for us. Similarly payment of dues to retired personnel must receive a very high priority. The **individual grievances** must be settled on **war footing** to avoid frustration and loss of efficiency. We have to anticipate the problems and needs of our people and take action well in time.

54. The people are our greatest asset. Each one has to be treated as a 'Gentleman' till he proves otherwise. We have to work through the strengths of each other rather than bringing weaknesses into focus all the time. We should undertake critical analysis of activities, ideas and performance of others **but not criticize.** We have to have some tolerance for the mistakes done in good faith. Mistakes as a result of callousness or due to reasons other than normal have to be dealt with severely. The annual appraisal of performance of our junior colleagues must be done objectively taking into account the potential and performance of the individual.

55. There has to be genuine love for each other and a quick response to the difficulties and grievances of the junior colleagues. Needless for me to add that honour of the organisation should be our first concern, always and every time. The welfare and well-being of our junior **colleagues** should come next. Our own comfort should come last. We should not let down the organisation even under worst provocation. **UNITED WE STAND DIVIDED WE FALL.**

Safety and Environment

56. **The safety and environment is another area of serious concern and merits a lot more attention than in the past.** Every development activity creates its own problems and damage to the environment. We should take remedial measures to maintain healthy environment. Tree plantation should receive even more priority. The sites which we abandon after drilling or seismic data acquisition should be left in such a manner that no one can find that there was any activity in the area. We have to make sure that the equipment for fire-fighting and blow-outs is available and is in serviceable condition. The prevention of fires and blow-outs is even more important. **You must read the Perspective Plan on Environmental Management (85/86 89/90). We are perhaps the only organisation in the country which has produced such a document on this important subject.**

57. It is our social obligation to interact with the people of areas in which we operate and assist them in development activities and through the ladies clubs render assistance in medical education and sanitation. We must provide for these social obligations through budgets.

Future Direction

58. ONGC is in the business of energy and we have to keep our options open for future in the event of a break through in energy area. You are aware that ONGC has been entrusted with an Insitu-coal gasification Project. We are also having a look at the geothermal energy. The field of solar energy should also attract our attention for assistance in the development of this energy source.

59. **We have to keep an eye on exploration in basins outside India which would enable us to have an access to the oil in foreign basins and have a window to the technology abroad, in addition to training and development of our personnel. Today perhaps is the time for acquisition of high technology companies abroad, so that we have presence in developed countries for easy access to technologies and to develop ourselves as an International Company.**

60. You have to take good care of your health and must devote some time to sports and cultural activities. As an organisation we have the moral duty to assist in the development of sports. The prestige of the country is enhanced by the performance of its sportsmen in the **International Meets** and is important for the high morale of the country. ONGC has taken a number of initiatives in this regard. It is our objective to have teams in hockey, football, volleyball, basket ball and cricket of national standard.

61. Please do not hesitate to respond to any of the issues referred to in this letter or anything else you may have in mind for the success of our organisation and its members.

68. I have very high expectations from each one of you and have no doubt that you would make every effort to act as owners of this organisation and bring out the desired efficiency, economy and excellent results.

With best wishes

Yours sincerely

(S.P. WAHI)

Appendix – N

The Cauvery Basin

Cauvery Basin, where geo-scientific surveys for oil exploration was started for the first time by ONGC in 1959, comprises a sedimentary basinal area of 27,000 sq. km. in the State of Tamil Nadu and the Union Territory of Pondicherry.

The first onland well Pattukottai-1 was drilled in 1963 and the first hydrocarbon strike followed closely in 1964, when Karaikal-1 showed the presence of hydrocarbons and later Karaikal-10 followed a small quantity of oil emulsion with low pressure.

The first offshore well in the Cauvery basin, Mannar IA, was drilled by Asamara (India) Ltd in 1979. Subsequently, ONGC resumed offshore drilling and had its first success offshore when PY-1-1 on Portonovo structure produced gas in 1980 followed by crude oil at PH-9-1, a well drilled in North Palk Bay in 1981.

Oil and gas was struck at the exploratory well Narimanam-1 in July, 1985. The well was spudded on March 28, '85 and drilled to the depth of 2321 metres using ONGC owned rig E-760. The well has potential to produce 250-350 barrels of oil per day.

The trial production from Narimanam-1 is of paramount significance as this basin gets upgraded to category-1 along with Bombay, Cambay and Assam, from where we have been producing oil for quite some time.

During the 7th Plan, ONGC plans to drill 36 wells, both onshore and offshore. Two onland rigs are presently in operation in this basin, at Tiruvarur and Kovilkallapal. Two more land rigs are proposed to be added and one offshore rig has also been planned during the years 1986-88. An investment of the order of Rs. 124 crores had so far been made in Cauvery basin till the Sixth Plan and a plan out lay of approx. Rs. 173 crores has been envisaged during the 7th Plan.

A dream come true

A 16-year-old dream has come true. Tamil Nadu has been put on the oil map of India with the commencement of trial production from Narimanam-I in the Cauvery Basin. The Narimanam crude has been found to be of very high quality, giving a new dimension to exploration in the region. This onshore production is as significant as the discovery at Gandhar-I in the Cambay Basin of Gujarat. The Cauvery Basin which has now been upgraded as "category one" along with Bombay, Cambay and Assam may be expected to come in for more intensive exploration in the next few years. The ONGC deserves a pat on the back for persisting with oil exploration in the Cauvery Basin and putting Narimanam on trial production in less than one year after it was found.

The ONGC commenced oil exploration in the Cauvery Basin in 1959. The basin comprises an area of 25,000 sq. km. in Tamil Nadu and Pondicherry and an offshore area of 23,000 sq. km. So far, 20 exploratory wells onland and 18 wells offshore have been drilled. Oil and gas were struck at the exploratory well in Narimanam-I in the coastal area of Thanjavur district in July last year. The well's potential has been estimated at 250 to 350 barrels per day. Initial production will be 20 tonnes per day. The crude is presently being transported by tankers to the Madras refinery for processing. Depending on further results, a pipeline will be laid to transport the crude to Madras. The prognosticated reserves in the Cauvery Basin, onland and offshore, are 370 million tonnes. The recoverable reserves by primary recovery are estimated at 70 to 80 million tonnes. Experts are of the view that the recoverable reserves can go upto 120 million tonnes by resort to secondary and tertiary recovery. Spurred by the success at Narimanam, the ONGC which is presently operating two onland rigs in the Cauvery Basin proposes to press into service two more rigs in the current year. Contingency plans for more rigs have also been drawn up. The ONGC plans to drill six deep wells onland and offshore during the Seventh Plan period. Its efforts in the Cauvery Basin will be supplemented by the Soviet Union which has agreed for the first time to participate in a big way in onland exploration in India. It has been allotted 7000 to 8000 sq. km. which represent approximately one-third of the Cauvery onland basin. India will welcome Soviet participation because the Soviet Union has agreed not to stake a claim for a share in any production that may result from its intensive exploration. Narimanam has opened up vistas of prosperity for the people in the coastal areas of Tanjavur district now dependent solely on agriculture. So far, Rs. 124 crores have been spent on oil exploration in the Cauvery Basin. The projected outlay during the Seventh Plan period is about Rs. 173 crores. This will have to be stepped up if a meaningful thrust is to be made in the wake of the success at Narimanam.

(Editorial, The Economic Times, Feb 25, '86)

Appendix – O

November, 1986

REPORTER

THE CORPORATE JOURNAL OF THE OIL AND NATURAL GAS COMMISSION

MINISTER ALL PRAISE FOR THE NEW VOCATIONAL CENTRE

The Union Minister of State for power, Mrs. Sushila Rohatgi, who also held the charge of Petroleum and Natural Gas, visited the Keshava Deva Malaviya Institute of Petroleum Exploration on October 3. On her arrival with Chairman Dr. S.P. Wahi, the Minister was received by Mr. S.N. Talukdar, Member (Exploration), Dr. S. Ramanathan, Member (Personnel) and Mr. K. N. Bhave, Director of the Institute. She was taken round the Museum and other departments of the Institute.

Later Mrs. Rohatgi had discussions with senior executives of ONGC. Dr. S.P. Wahi apprised her of ONGC's interaction with universities and different research institutes of the Commission on fundamental and applied research. The Minister appreciated the efforts of ONGC in this direction, particularly in converting the R&D units to profit centres. She congratulated the organisation for its initiative in concentrating on high technology areas and said that this should be a path finder to other public sectors in the country.

The next day in the morning, the Union Minister inaugurated ONGC's Vocational Centre and Priyadarshini Gas Sewa run by the Ladies Club.

Mrs. Rohatgi congratulated the members of Ladies Club for starting such vocational centres not only in Dehra Dun but also in other places. As a leading public sector, ONGC was fulfilling its social commitments by taking all these initiatives, she observed. The Minister commended the performance of ONGC and said it was a matter of pride for it to be the 12th profit making giant in the world.

Earlier, in the welcome address Mrs. Shobhana Wahi, patron of the Ladies Club referred to the various activities of the Club aimed at uplifting the status of women and promoting national integration.

Also prsent on the occasion was Dr. S. Ramanathan, Member (Personnel).

Presently, a large number of jobs viz., printing, shredding of obsolete records, sewing, knitting, masala preparation etc. are being undertaken by the Vocational Centre. Priydarshini Gas Sewa is handling the supply of LPG to ONGC employees, numbering over 1000, residing in the ONGC Colony.

Mr.S.P. Dhir, formerly Senior Director and Consultant to Secretariat, is the Supervisor of the Vocational Centre. Most of the employees in the Centre are widows, or dependents of deceased employees of ONGC.

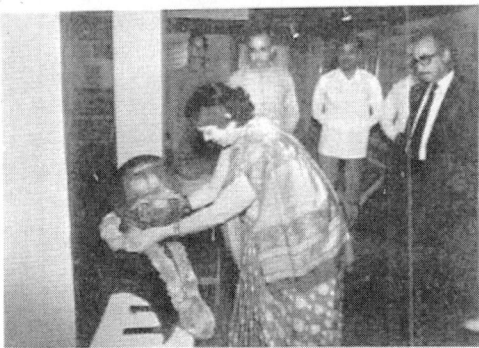

Mrs Rohatgi at the KDMIPE Museum

Handing over the first gas cylinder on the inauguration of Priyadarshini Gas Sewa.

Appendix – P

ONGC CONTRIBUTES TOWARDS BEAUTIFICATION OF DEHRA DUN

As part of its social commitment to improve the conditions of towns near its work centres, the Oil and Natural Gas Commission has donated Rs. 2.50 lakhs to the Dehra Dun Municipality for providing sodium light fittings in the main areas of the city. It has been suggested that the lights be fitted along the Rajpur Road and Kaulagarh Road, which are mostly used by the ONGC employees.

At a simple function organised in Tel Bhawan on March 16, 1985, Chairman Col. S.P. Wahi presented the cheque to Mr. Atul Chaturvedi, Administrator, Dehra Dun Municipality.

It may be recalled that earlier the Commission had contributed Rs. 1.78 lakhs towards the renovation of and seating arrangements for the Dehra Dun Town Hall. The Commission has also made a contribution of Rs. 50,000 towards the Santosh Trophy matches at Dehra Dun, to promote sports activity in the city.

Chairman seen with Mr. Atul Chaturvedi, District Magistrate. Standing close to Col. Bhem is Mr. Shailendra Singh, Superintendent of Police. On the right of Mr. Chaturvedi is Mr. Premendra Singh Verma, Officer-in-charge, Dehra Dun Municipality Board

Appendix – Q

OAPEC DELEGATION VISITS DEHRA DUN

Before the commencement of the seminar at Delhi, Dr. Ali Ahmed Attiga visited the Oil & Natural Gas Commission establishments at Dehra Dun on February 14. He was impressed with the Institute of Drilling Technology and KDM Institute of Petroleum Exploration and said that ONGC within short span of time has built up training and R&D facilities of

Dr. Attiga seen with the Chairman and Members at Tel Bhawan

productivity in all aspects of its exploration. Chairman, Dr. S.P. Wahi explained to the Secretary General the new strategies being evolved and future thrust areas like exploration in deeper waters, development of new technologies, optimisation of design of offshore structures to reduce costs, upgradation of resources by accelerating the process of delineation

through unconventional methods etc. He informed the functioning of core groups at the corporate and project levels to constantly monitor cost reduction measures and suggest improvement in productivity.

In honour of the visiting dignitaries, a cultural programme was organised at the AMN Ghosh Auditorium. After the conclusion of the Delhi seminar, the members of OAPEC de-

Dr. Attiga inaugurating the cultural programme

international standards. After having listened to the corporate presentation on the growth and prospects of ONGC, Dr. Attiga spoke in glowing terms of the efforts of ONGC to improve

Members of the OAPEC delegation at the Institute of Drilling Technology

In one of the laboratories of KDMIPE

legation also visited ONGC, Dehra Dun and were apprised of the various Research and Training programmes being conducted here. The delegates felt that the occasion should not be the end of a process, but mark the beginning of an era of mutually profitable cooperation between the petroleum exporting countries of the Arab and ONGC.

Appendix – R

ECONOMY & BUSINESS

ONGC
THE COLONEL'S 2-FRONT WAR

The No.1 profit-maker in India's anaemic public sector, the Oil & Natural Gas Commission (ONGC) is paradoxically not without its financial and other travails. Right now, ONGC's charismatic chairman Colonel S.P. Wahi (59) — whose extended tenure of an unprecedented seven years ends November next — is battling with the Union government on two fronts. First, he wants the government to pay ONGC at least Rs.300 more per tonne of crude. Second, he has declared a firm nyet to the proposal that ONGC hand over its natural gas processing units (at Hazira and Uran) and gas prospecting projects (inter alia at Tripura, the Krishna-Godavari delta and western onshore) to the Gas Authority of India Ltd. (GAIL).

As far as the price of crude is concerned, Wahi points out that this is perhaps the only commodity in India of which the domestic price is well below the international average. As a matter of fact, indigenous crude oil prices were based on import parity considerations till July 1981, when an administered pricing mechanism was introduced. The latter broke up the price of crude oil into two parts: a base price (paid to ONGC) and statutory charges like royalty and cess (paid to state and Central governments respectively). Since 1981, crude prices have been raised thrice — but on each occasion, the beneficiaries have been the state and Central governments. In other words, the base price has been left unchanged, while cess and royalty have been upped.

As of date, ONGC's net realisable base price is Rs. 1,021 per tonne, while a royalty of Rs. 192 per tonne (cf. Rs. 61 in 1981) and a cess of Rs. 600 (cf. Rs. 100) per tonne makes up the ruling aggregate price of Rs. 1,813 per tonne. Incidentally, there is no correlation between this indigenous crude oil price and the domestic prices of petroleum products since the downstream refineries follow a different pricing mechanism based on retention prices.

Legitimate grouse. Wahi's grouse — which is legitimate — is that while the government has merrily upped the prices of end-products like petrol, diesel and kerosene, and has also increased cess and royalty to augment its own income, the actual producer (ONGC) has been left to tighten its belt.

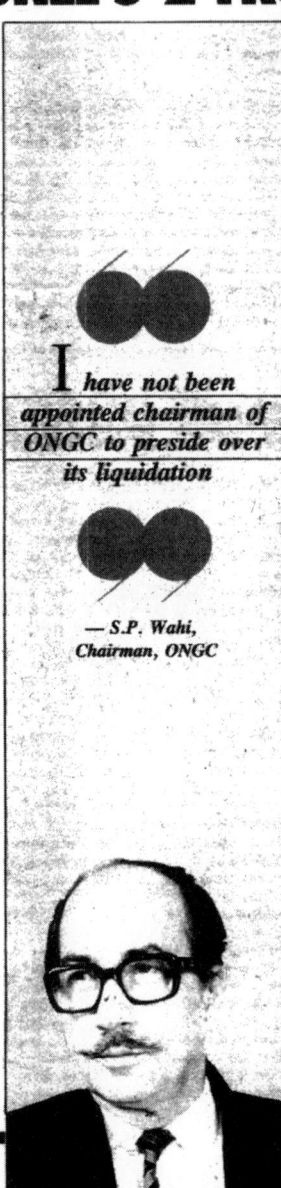

I have not been appointed chairman of ONGC to preside over its liquidation

— S.P. Wahi, Chairman, ONGC

"Worse still, everybody thinks these price hikes are fuelling ONGC's profits — a notion which could not be further from the truth," he rues.

In reality, says Wahi, only 13%-15% of ONGC's total cost of production is "controllable," while the balance 85%-87% is "not controllable, being accounted for by statutory costs, financing costs and such like." To illustrate, ONGC shells out at least Rs. 120 crores every year by way of a subsidy bill on account of indigenising services, contracts and materials. For instance, all Indian suppliers of equipment and services have to be given a price preference of as much as 40%. Urges Wahi: "The government should compensate ONGC for being forced to pay these subsidies."

He argues that the ONGC makes the kind of profits it does (estimated at around Rs. 1,600 crores for the year ended 31st March 1988) thanks to increased productivity all round. The statistics reeled off are impressive. In 1986-87, crude oil production per employee was 643 tonnes (cf. 327 tonnes in 1980-81), gas sales were 116.5 thousand cu.m (cf. 34.6 thousand cu.m) per employee while the cycle speed in drilling stood at 705 meters per rig month (cf. 475 metres). "Our increasing profitability is due to higher productivity, which has become a culture at ONGC. Surely, our efficiency should be rewarded and not punished," expostulates Wahi.

However, the Union government does not seem to be in a particularly generous mood. Far from it. A few weeks ago, a bombshell landed on Wahi's desk in the form of an innocuous demi-official (DO) letter written to him by B.S. Lamba, a joint secretary in the ministry of petroleum and natural gas (P&NG). Lamba "advised" Wahi to hand over a substantial chunk of ONGC's assets pertaining to natural gas development projects and processing facilities to GAIL. To rub salt into the wound, as it were, he ordained a cut-off date of 31st March by which the transfer ought to be effected. The financial angle, promised Lamba, would be sorted out later.

Feisty reaction. Predictably, Wahi has reacted with his typical feistiness. The "cut-off date" has come and gone, and ONGC has not parted with any assets yet. "I have not been appointed chair-

ECONOMY & BUSINESS

man of ONGC to preside over its liquidation," he says, echoing Winston Churchill. Adds Wahi: "GAIL wants to have a baby without any labour pains. Our boys have built up these assets by their sweat and toil. Why should we gift them off? Besides, since when has the government decided to communicate important policy decisions by way of DO letters?"

The two schools of thought espouse opposite views. The GAIL lobby (headed by GAIL's ex-IAS chairman V. Nayyar and championed by Lamba) pleads that GAIL should oversee the entire spectrum of processing, transportation and marketing of natural gas, which is fast emerging as an attractive substitute for oil. The converse argument is that exploitation of the country's natural gas resources (along with crude) is part of ONGC's mandate. Therefore GAIL should restrict itself to utilising the gas, perhaps by working towards the creation of a national gas grid.

It is anybody's guess as to how the ONGC-GAIL tussle will finally end. At the moment, the vital post of secretary, P&NG, is vacant, with coal secretary S. Varadan merely holding additional charge.

Even as the tussle is fought in and over the P&NG departmental files, Wahi has pulled an innovative ace from his sleeve. In an astute move aimed at "further broadbasing the managerial decision-making boundaries at ONGC" — but which will inevitably also enable the commission to win the support of influential constituencies — several advisory committees have been set up. These cover fields such as exploration, human resource development and management. The first two have already met, while the advisory committee on management is scheduled to have its maiden meeting at Dehra Doon on 14th May. Not surprisingly, two leading lights of the media (Dr. Ram Tarneja of *The Times of India* and Ashok Advani of *Business India*) have been invited to sit on the committee.

Clearly, our regressive public sector culture does not accept a healthily-in-the-black bottomline as proof enough of work well done. As such, ONGC still needs to win friends and influence people. And should Wahi turn out to be the gallant loser in the looming price and natural gas battles with his recalcitrant government bosses, there will certainly be no dearth of admiring sympathisers to provide him comfort.

■ *Sujoy Gupta*

Appendix – S-1

ADVISORY COUNCIL ON EXPLORATION STRATEGY

The first meeting of the Chairman's Advisory Council on Exploration Strategy (CACES) which was held at the KDM Institute of Petroleum Exploration, Dehra Dun on February 20-21, 1987, has brought about a greater element of refinement in the policy relating to the exploration strategy and identified a number of thrust areas.

On behalf of the Chairman, Mr. P.K. Chandra, Member (Exploration) welcomed the delegates and sought their candid advice on the present exploration strategy of the organisation. He said a dispassionate view of the explorationists and experts assembled in the meet would help ONGC to adopt corrective measures. Since most of the participants of the Advisory Council have had long association with ONGC, their suggestions would be very valuable, Mr. Chandra added.

In order to acquaint the members of the CACES on various on going projects and strategies, a number of presentations were organised. The first to present was Dr. B.I. Sharma, General Manager (Overseas Operations). Describing the efforts of the developed countries like France, Italy, Norway and Japan to set up their national oil companies, Dr. Sharma went on to plead that in addition to indigenous exploration, successful oil companies undertake exploration ventures in promising basins on a global basis. Inspite of drilling many dry wells, major multi-national and national oil companies are still exhibiting active interest in Chinese, Thai, Tanzanian, Vietnamese and Indian sedimentary basins. As India is a net importer of oil and also not blessed with prolific petroliferous basins, the Govt. of India would do well to dictate a policy, which apart from indigenous exploration also encourages oil exploration ventures abroad to meet its long term energy needs and objectives. Dr. Sharma highlighted ONGC's rich experience of operating overseas and spelled out some of the strategies which could be adopted by it in this direction.

This was followed by a presentation on exploration strategies by ONGC. Mr. K.N. Murty gave a detailed analysis of ONGC's activities.

The exploration strategy presently being followed is to intensify exploratory efforts in the producing basins of Bombay Offshore, Cambay Basin and Upper Assam and to intensify efforts in other basins like Rajasthan, West Bengal, Assam-Arakan fold belt, Krishna-Godavari, Cauvery, Andamans, Himalayan Foothills, Tripura etc. In line with this objective, exploratory efforts have been going on, in addition to the producing basins, in Rajasthan, Ganga Valley, Krishna-Godavari, Cauvery, West Bengal onland, Tripura and Assam Arakan fold belt. Exploration has restarted after assessing and evaluating the seismic data in Andaman offshore and recently in Himachal Predesh and West Bengal offshore.

Mr. K.N. Bhave made a presentation on Frontier Technology to Solve Exploration Challenges in 80s and 90s. The exploration methods are undergoing revolutionary changes, thanks to the rapid advancements in the electrical, electronics and optical technology and the development of new precision techniques of chemical and physical analysis. Mr. Bhave maintained that this has to be so, as the easy-to-discover structural fields have

Members of CACES seen in one of the laboratories of KDMIPE

mostly been discovered and we have entered into an era of exploration for subtle stratigraphic traps and complex structures, and in areas of hostile terrain, poor seismic reflections, high pressure and other drilling problems. The exploration discipline of ONGC has not lagged much behind in introducing the new techniques once these have been tested and found acceptable in the advanced countries.

This important and interesting discussion was also greatly contributed by Dr. S. Ramanathan, Member (Personnel) who said the growth of geological reserves established by ONGC in the last six years has been significant. A multi-pronged and aggressive exploration strategy has been adopted to accelerate the upgradation of reserves

Appendix – S-2

conference with ONGC scientists

and optimum mix of low risk and low reward in known basins and high risk & probables in high reward basins. The achievement of self-reliance in oil and gas is related to the extent to which we are able to establish exploitable hydrocarbon resources. The setting up of this Advisory Council by the Chairman, ONGC, will give further fillip to this effort.

Lively discussions ensued the next day. Chairman Dr. S.P. Wahi who steered the discussion, said that ONGC scientists have now earned international reputations for their exploration strategies. The discovery of Gandhar and several offshore structures have taken the ONGC's credibility very high. Yet, we must strengthen the techno-economic groups functioning in each region and every decision should be arrived on pure commercial basis. Endorsing the view of eminent members of CACES, the Chairman said that cost-consciousness should form a part of our exploration and drilling decisions and each of the geoscientists should undertake his activity in a spirit of dedication.

There was a presentation on ONGC's activities in Himalayan foothills. Mr. P.V. Krishnan, DGM, highlighted the new approach adopted to drill the deep wells in the area.

Dr. Vijay Kelkar, Former Advisor (Economic Planning Policy), Department of Petroleum and now Chairman, Bureau of Industrial Costs and Prices, Mr P.K. Dave, Former Secretary to Govt. of India, Mr Lovraj Kumar, former Secretary, Department of Petroleum and now Advisor in the Advisory Board on Energy, Dr. Hari Narain, Chairman of the Scientific Advisory Committee to the Petroleum Ministry, Mr S.N. Talukdar, former Member (Exploration), ONGC, Dr. R.K. Pachauri, Director, Tata Energy Research Institute, Mr B.S. Negi, former Chairman, ONGC, Mr. A.B. Das Gupta, former Chairman, Oil India Limited, Mr C.K.R. Sastri, Mr S.N. Sengupta, former General Manager, ONGC and the senior geoscientists in the Commission participated in the conference.

The Ladies Club organised a cultural programme in the evening in the honour of the members of the Advisory Council.

Appendix – T-1

Prime Minister lauds the efforts of ONGC

The South Bassein BPA Platform complex of ONGC was formally dedicated to the nation by the Prime Minister, Mr. Rajiv Gandhi on February 16, 1989 at a simple ceremony aboard the platform at Bombay Offshore.

The Prime Minister lauded the efforts of ONGC Scientists, Engineers and Technicians in commissioning the giant South Bassein complex. (See message). He said ONGC has been carrying out a very vital task for the nation. Any nation today is judged by the amount of energy it consumes, "India during these past few years has increased its energy consumption tremendously. ONGC has been greatly responsible for the energy that we have been using. ONGC has also demonstrated that public sector can be efficient and can compete with the best anywhere on the globe. I would like to congratulate, the officers, the men, everybody in ONGC for the performance you have shown and the proud way you have held our flag high".

Earlier on the Minister of State for Petroleum and Natural Gas, Mr. Brahm Dutt, welcomed the Prime Minister and said it was due to the vision of Pandit Jawahar Lal Nehru that India could initially organise efforts in exploration. It was during Mrs. Indira Gandhi's leadership that the oil exploration programme matured and progress was made and now that Mr. Rajiv Gandhi is at the helm we have entered the gas era for economic prosperity." The Chairman, ONGC, Col. Wahi thanking the Prime Minister said that even though the visit was brief, this would be a big morale booster for the employees of the ONGC. "ONGC is conscious of its responsibilities to meet the growing energy needs required to fuel the economic development of the country and that ONGC shall relentlessly work towards this goal." Speaking about the production scenario of Bombay High, the Chairman said that presently Bombay High meets 60% of the oil production of ONGC. Today ONGC produces 30 MT per annum whereas by 1994-95, ONGC would be producing 45 MT and Bombay High field is expected to contribute 40% of this production. A lot more oil and gas is yet to be located.

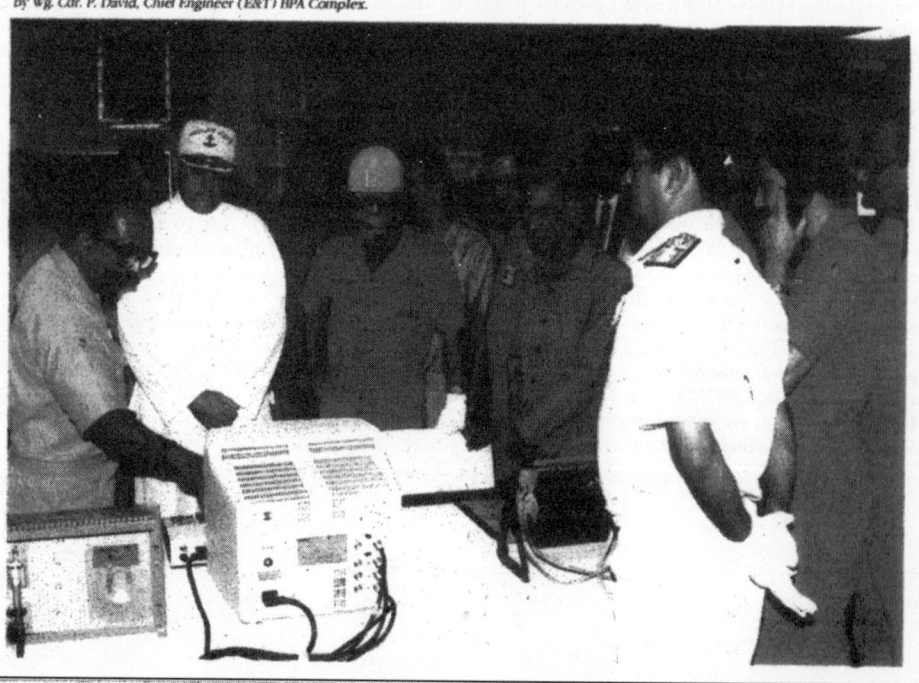

The Prime Minister being explained the operations of Titan Communication & Computer network by Wg. Cdr. P. David, Chief Engineer (E&T) BPA Complex.

Appendix – T-2

Exploration is being extended to deeper waters he further added.

South Bassein is one of the giant gas fields in the world with reserves of 395 MT of oil and oil equivalent of gas. The gas is routed to the Hazira onshore terminal via a 36" diameter pipeline which has a handling capacity of 20 Million cubic metres per day. Built to meet the stringent requirements of sour gas transportation, the pipeline is not only India's longest subsea pipeline but also one of the heaviest and largest diametre pipelines in the world.

The Prime Minister was accompanied by Mr. Brahm Dutt, Chief Minister of Maharashtra, Mr. Sharad Pawar, Secretary Petroleum and Natural Gas, Mr. H.K. Khan and Chief of Naval Staff, Admiral Nadkarni.

A group photograph with the Prime Minister.

PRIME MINISTER

MESSAGE

The Oil and Natural Gas Commission has been carrying out a very vital task for the nation. Any nation today is judged by the amount of energy it consumes. India during these past few years has increased its energy consumption tremendously. The ONGC has been largely responsible for the energy that we have been using. The ONGC has also demonstrated that the public sector can be efficient.

The public sector can compete with the best anywhere on the globe and I would like to congratulate the ONGC for its performance.

New Delhi
February 24, 1989

Appendix – U

ONGC'S STRATEGY FOR ENVIRONMENTAL PROTECTION

ONGC's operations fall primarily in three categories: geo-scientific surveys, exploratory and development drilling and production operations. Drilling and production activities inevitably interact with the environment through short and long term physical, chemical and biological changes. That these changes can bring about environmental impact on the quality of air, water and land had been visualised long back. For the first time in the history of oil industry, a 5 year environment management plan was drawn up by the ONGC, titled "Perspective Plan on Environmental Management" early in 1985. This comprehensive document was preceded by another educational booklet by ONGC's Safety & Environment Management discipline, emphasising the role of all employees in the preservation of the environment around them. Of more far reaching importance was the Resolution adopted by the Commission as far back as in 1983, setting out its policy on Environment Management. The aim of all these was the evolution of a technology in the design and construction facilities, preservation of flora, fauna and aquatic life; regenerating forest cover and rationalising land use wherever the ONGC is operating. In the 5-year Environment Management Plan, undertaking environmental impact assessment before project commencement in mandatory.

ANTI-POLLUTION MEASURES

In implementing the Plan, at onshore drill sites, the waste pits are being brick or polythene-lined and enclosed in high ring bunds because overflow or seepage of fluids from the pits can contaminate stallation, the separated formation water with its oil content often contaminate water sources unless it is effectively treated prior to discharge. It is therefore being treated in Effluent Treatment Plants (ETPs) to accountable standards prior to discharge. In some areas, the effluent is vaporised in specially designed flare pits. ETPs have been installed at Kalol, Jhalora, Sanand, Nawagam, Sobhasan and Santhal in Gujarat; Lakwa in Assam and Uran in Maharastra

Environmental Conservation

In order to minimise any adverse impact on the environment surrounding a new project, Environmental Impact Assessment is invariably undertaken and clearances obtained. The services of well-known consultants and experts are drawn upon by ONGC, wherever necessary. Tree-plantation at work-sites, offices and residential complexes have become a regular feature. Plant survival is constantly monitored to ensure sustained growth and the terms of contracts are on the basis of the number of plants that survive. In consonance with Government's directive, fresh guidelines have been issued. Financial support has been provided to various State Departments and local agencies for development of parks and sanctuaries in and around ONGC complexes.

In pursuance of ONGC's policy, selected personnel have been exposed to specialised training in India and abroad, Audio-visual programmes, poster campaigns and "Safety and Environment Week" are organised to promote environmental awareness.

Air Pollution Control

Air pollution is now minimised by gradual reduction of gas flaring, as more and more facilities are created for the use of gas in the production of LPG, fertilizers and power generation. Where flaring is inevitable, it is rendered smokeless by steam injection. Since radiant heat of flares can be harmful to fauna and flora, the flare is suitably enclosed. The ONGC also takes care to prevent pollution due to sulphur dioxide gases. Gas sweetening and sulphur recovery plants wherever required are being set up likewise, near ONGC's process plants, ambient air quality is regularly monitored.

Marine Pollution Control

In the marine environment, uncontrolled flow of oil from a well or a production platform might result in pollution of sea. Offshore production wells are therefore invariably fitted with surface-controlled sub-surface safety valves, limiting oil spills into the sea. Under-water pipelines are provided with automatic shut-off valves which close in case of rupture of a pipeline, thereby reducing leakage of oil into the sea. Effluents from production platforms are discharged into the marine environment only after adequate prior treatment in accordance with the prescribed standards. The contingency of likely oil-spills has been assessed and adequate equipment and facilities acquired and installed.

Research and Development

In order to cope with the pace of technological development, the R&D aspects of environmental conservati through in-house facilities as well as outside agencies continue to be geared up.

Land is being acquired to house ONGC's Institute of Petroleum Safety & Environment Management. The Institute will offer training courses in safety and pollution control, in safety and environment management and R&D on safety and environmental problems.

Environment Management in ONGC is a way of life and a pointer for all industrial establishments in the country.